# RELIABILITY IN THE ACQUISITIONS PROCESS

# LECTURE NOTES IN STATISTICS

A Series Edited By

D. B. OWEN

Department of Statistics
Southern Methodist University
Dallas, Texas

Vol. 1 Incomplete Block Designs, *by Peter W. M. John*
Vol. 2 Matrix Derivatives, *by Gerald S. Rogers*
Vol. 3 Statistical Analysis of Weather Modification Experiments, *edited by Edward J. Wegman and Douglas J. DePriest*
Vol. 4 Reliability in the Acquisitions Process, *edited by Douglas J. DePriest and Robert L. Launer*

*Additional Volumes in Preparation*

# RELIABILITY IN THE ACQUISITIONS PROCESS

edited by

Douglas J. DePriest
Statistics and Probability Program
Office of Naval Research
Arlington, Virginia

Robert L. Launer
Mathematics Division
Army Research Office
Research Triangle Park, North Carolina

MARCEL DEKKER, INC. New York and Basel

Library of Congress Cataloging in Publication Data

Main entry under title:

Reliability in the acquisitions process.

(Lecture notes in statistics ; v. 4)
Includes indexes.
1. Reliability (Engineering)--Congresses. I. DePriest, Douglas J., II. Launer, Robert L.
III. Series: Lecture notes in statistics (Marcel Dekker, Inc.) ; v. 4.
TA169.R44 1983 620'.00452 82-22082
ISBN 0-8247-1792-9



MARCEL DEKKER, INC.
270 Madison Avenue, New York, New York 10016

Current printing (last digit):
10 9 8 7 6 5 4 3 2 1

PRINTED IN THE UNITED STATES OF AMERICA

# Preface

Reliability theory has been and continues to be an active area of research. This volume contains the invited papers presented at the combined Office of Naval Research (ONR) and Army Research Office (ARO) Reliability Workshop held at the Department of Commerce, Departmental Auditorium in Washington, D.C. on 29 April through 1 May 1981. The objectives of this "state-of-the-art" workshop were to review recent developments in reliability research and to provide a forum to facilitate communications between DoD personnel and academic researchers.

This book is not intended to be a text, but rather to serve as a valuable reference source. It is a compilation of recent research results by some of the leading researchers in reliability. The invited speakers were requested to focus their papers around the theme "Reliability in the Acquisitions Process." The book is divided into three parts. The first part consists of two keynote papers. The first paper by Dr. Thomas Varley, Director of the Operations Research Program, ONR, focuses on concepts and pertinent issues related to the acquisitions process. The other paper by Professor Walter Smith, Chairman, Department of Statistics, University of North Carolina, develops some interesting features of a class of stochastic processes which have been useful in renewal theory applications. The second part contains papers on various aspects of system reliability which was the major emphasis of the ARO portion of the workshop. The third part represents a juxtaposition of widely different approaches and concepts in reliability; from applications of recent results for improvement or revision of military standards to discussion of abstract theoretical methodology; from an exposition on a particular failure rate model to nonparametric estimation subject to reliability growth restrictions.

The theme was selected to stimulate interest and increase the level of awareness of academic researchers in the acquisitions process. The military acquisitions process is the whole process of acquiring weapons systems and military material to meet the operational needs of the DoD. It is a sequence of programmed activities that result in decisions on system development, procurement and service use. Reliability and maintainability play an important part in the acquisitions process. With increased emphasis being placed on acquisition in DoD, it seems fitting that academicians should reflect on how their reliability research can affect our national defense.

Many people deserve to be acknowledged for their contributions to this volume. The joint workshop concept arose from discussions between Dr. Edward J. Wegman, Director of the Statistics and Probability Program at ONR and Dr. Robert Launer, Mathematics Division at ARO, during the 1980 Joint Statistical Meetings in Houston, Texas. We especially wish to thank Dr. Wegman for his help, constructive suggestions, and participation. Professor Nozer Singpurwalla (George Washington University) was most instrumental in coordinating local arrangements during the initial planning. Finally, we are grateful to all the participants and contributors, for without them this volume would not be possible.

DOUGLAS J. DePRIEST

ROBERT L. LAUNER

# Contents

# Contributors

HAROLD ASCHER, Management Information Division, Naval Research Laboratory, Washington, DC

G.K. BHATTACHARYYA, Department of Statistics, University of Wisconsin, Madison, Wisconsin

HENRY W. BLOCK, Department of Mathermatics, University of Pittsburgh, Pittsburgh, Pennsylvania

RICHARD DYKSTRA, Department of Statistics, University of Missouri—Columbia, Columbia, Missouri

HARRY FEINGOLD, Computation, Mathematics and Logistics Department, David W. Taylor Naval Ship Research and Development Center, Bethesda, Maryland

CAROL FELTZ, Department of Statistics, University of Missouri—Columbia, Columbia, Missouri

JOSEPH C. GARDINER, Department of Statistics and Probability, Michigan State University, East Lansing, Michigan

BERNARD HARRIS, Department of Statistics, University of Wisconsin, Madison, Wisconsin

M. VERNON JOHNS, Jr., Department of Statistics, Stanford University, Stanford, California

NORMAN L. JOHNSON, Department of Statistics, University of North Carolina, Chapel Hill, North Carolina

RICHARD A. JOHNSON, Chairman, Department of Statistics, University of Wisconsin, Madison, Wisconsin

SAMUEL KOTZ, Department of Management Science and Statistics, University of Maryland, College Park, Maryland

JANET MYHRE, The Institute of Decision Science for Business and Public Policy, The Claremont College, Claremont, California

DONALD B. OWEN, Department of Statistics, Southern Methodist University, Dallas, Texas

SAM C. SAUNDERS, Department of Pure and Applied Mathematics, Washington State University, Pullman, Washington

THOMAS H. SAVITS, Department of Statistics, University of Pittsburgh, Pittsburgh, Pennsylvania

NOZER D. SINGPURWALLA, Department of Operations Research, The George Washington University, Washington, DC

WALTER L. SMITH, Department of Statistics, University of North Carolina at Chapel Hill

ANDREW P. SOMS, Department of Statistics, University of Wisconsin, Madison, Wisconsin

THOMAS C. VARLEY, Office of Naval Research, Arlington, Virginia

ALAN WINTERBOTTOM, Department of Mathematics, City University, London, England

# RELIABILITY
# IN THE ACQUISITIONS PROCESS

# Part I: Keynote Papers

# Reliability and Maintainability in the Acquisition Process

Thomas C. Varley

*Office of Naval Research*

**Abstract**

This paper describes a concept for acquisition research as performed by the Navy, and describes the framework for including R&M in the research process. Included are concepts for understanding how R&M relates to the affordability issue and to improving current force capabilities.

## INTRODUCTION

There is an increasing concern as to the high cost of producing and maintaining weapon systems. Current trends indicate that these forces are all moving in the wrong direction. Complexity of systems are increasing, without improved maintainability considerations to lower the operating costs; unit production costs are increasing with ever increasing maintenance personnel skill levels; availability of skilled personnel is decreasing through lower retention of mid-career personnel, while manpower availability is decreasing due to past national demographics. Added to these individual trends is the interaction and relationships they combine to produce. We have shortfalls in personnel and equipment, and we have increasing operating costs which reduce our capability to create new assets under constrained budgets. We have reduced assets but they are costing us more. Where does research in reliability and maintainability interact with these trends and what can they do to reverse these trends?

## THE TRENDS

The role of R&M in the acquisition process is very broad and can have a significant impact on the long run operational capability of Naval forces. One generally thinks of R&M considerations as a method for improving equipment operation, but because of the interactions between all components of the system, improving equipment/system R&M has far greater implications.

Consider some of the following statistics: The availability of manpower for the remainder of this century will be 10-25% lower than we had in 1976. Figure 1 displays the percentage change over that 20-year time period. It is interesting that the same general picture is true for females, as well as males in the 17-19 year age range. With personnel available, and a greater demand being placed on this population by industry and academic institutions, the Navy must develop ways to decrease its personnel requirements. R&M has potential for doing this.

Figure 2 describes a second consideration—the decreasing number of Naval aircraft purchases over the last 10 years and their average costs. In 1970, we purchased 134 aircraft for 442 million dollars, while in 1980, we purchased 39 aircraft for 1,365 million dollars—a 400% decrease in aircraft; yet, over a 300% increase in procurement costs. Associated with this is an annual maintenance cost that averages about 6% of its acquisition cost per year per aircraft, a second area where R&M can play a significant role.

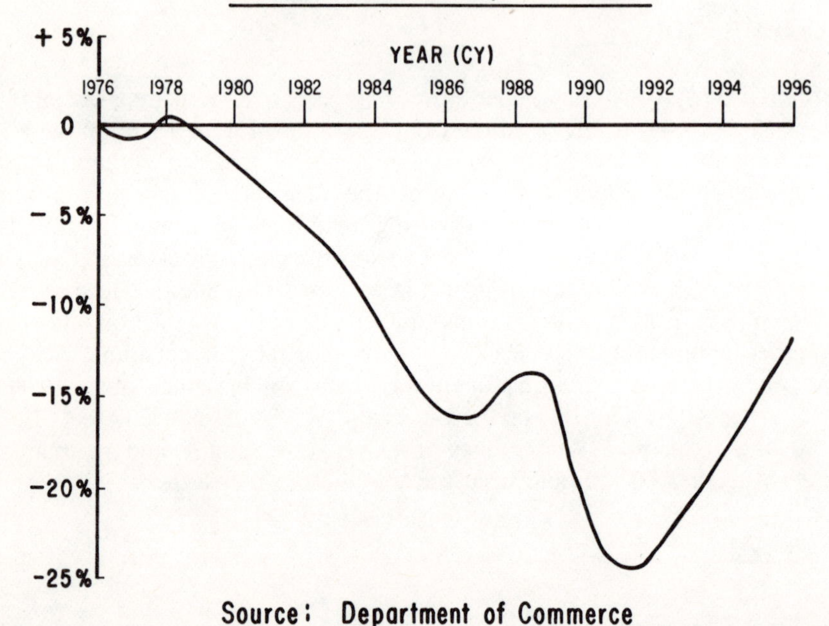


Figure 1

Declining Navy Aircraft Purchases

| Fiscal year | Fighters and attack planes bought | Average cost of each plane, million |
|---|---|---|
| 1970 | 134 | 3.3 |
| 1971 | 110 | 8.0 |
| 1972 | 114 | 7.5 |
| 1973 | 147 | 5.5 |
| 1974 | 149 | 6.4 |
| 1975 | 92 | 9.2 |
| 1976 | 77 | 10.6 |
| Transition quarter | 15 | 7.3 |
| 1977 | 93 | 9.7 |
| 1978 | 68 | 14.1 |
| 1979 | 69 | 20.2 |
| 1980 | 39 | 35.0 |
| Total | 1107 | |

**Figure 2**

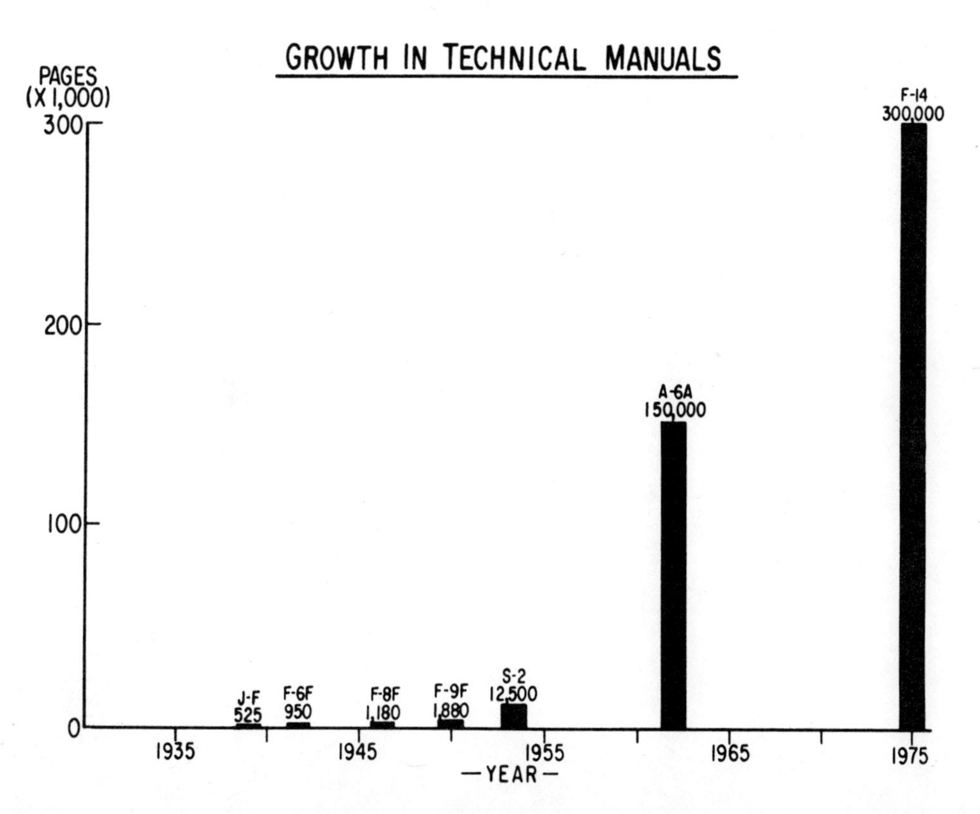


**Figure 3**

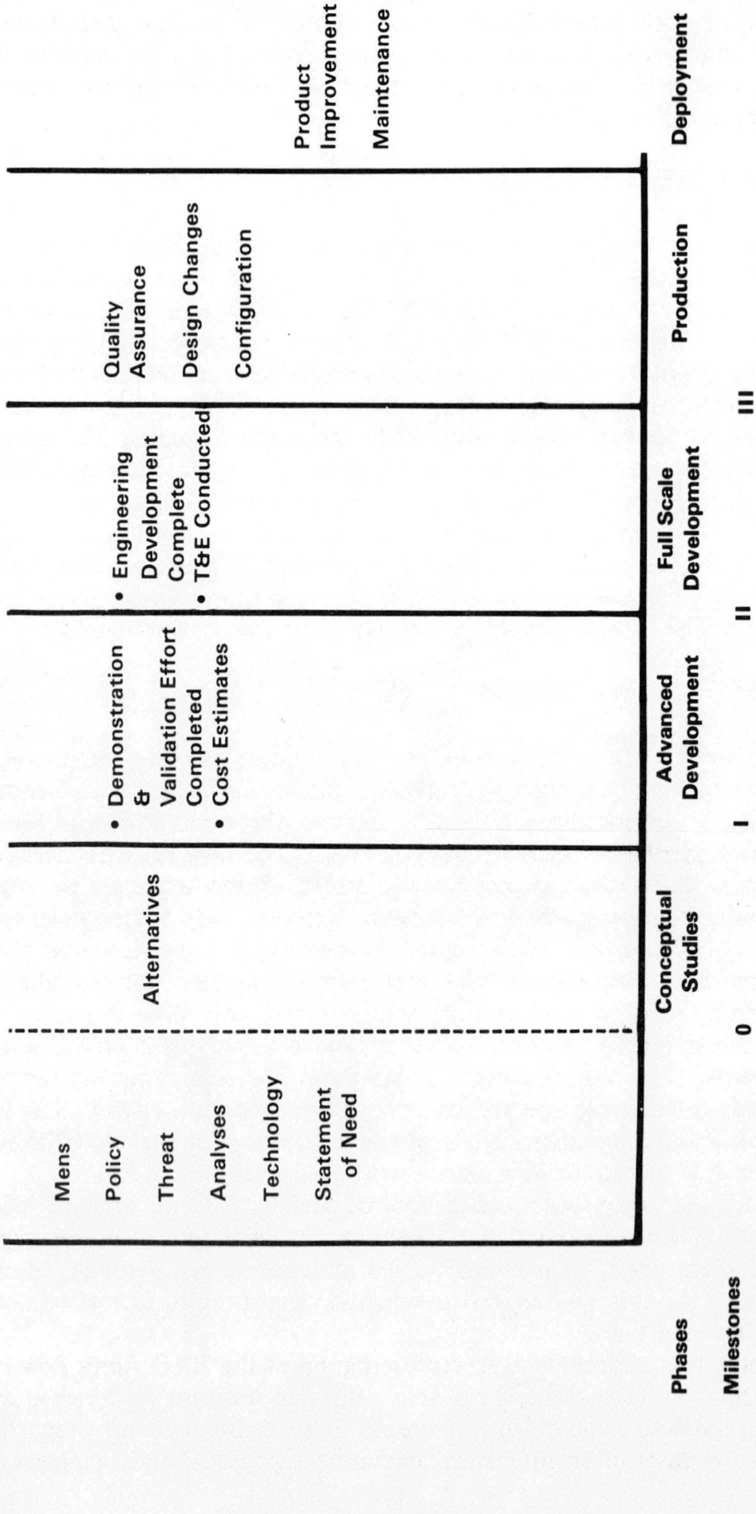

Mens
Policy
Threat
Analyses
Technology
Statement of Need
Alternatives
• Demonstration & Validation Effort Completed
• Cost Estimates
• Engineering Development Complete
• T&E Conducted
Quality Assurance
Design Changes
Configuration
Product Improvement
Maintenance
Phases
Conceptual Studies
Advanced Development
Full Scale Development
Production
Deployment
Milestones
0
I
II
III

Figure 4

The increasing complexity of aircraft is well illustrated by Technical Manual trends. Figure 3 shows the growth of Technical Manual size over the last 10 years. This growth is unavoidable in order to provide good documentation, but the control, updating, and training required is a considerable burden on Naval resources. This is another area where R&M can play a significant role in alleviating present difficulties.

## WEAPONS LIFE CYCLE

So far, we have been describing some particular areas where improving R&M could be beneficial. How do we take advantage of these potential benefits? First, we have to understand the cycle, the long life cycle of a major system, before we can understand the role R&M plays in the acquisition cycle. Figure 4 defines the various phases of the acquisition process from definition of the need through deployment of the system. To time phase Figure 4, the period to deployment is between 10-14 years while the initial operating life varies from 15-30 years for aircraft and ships, respectively. Together, the acquisition cycle and the deployment or operating life, represent the life cycle of the system. The critical decision point is Milestone III which is the decision to produce the system. Once the decision is made for production, future resources need to be allocated for its operation. These resources are of a time-phased nature and can be considered as "flows" through the system over a multi-year period.

## ACQUISITION FUNDS FLOW

Figure 5 is one attempt to describe the Navy budget in a very aggregated form with three major categories: Procurement funds, R&D funds, and something called 'cost of current assets.' These categories represent the flow of funds into the primary functional areas of Navy activity. The flow of funds through the procurement chain result in accumulated assets. These assets are the results of the acquisition process, which has a major decision node on the determination of force characteristics. R&D funds flow two ways; some flow into the force characteristics decision node and are used to develop various alternatives characteristics, and to produce prototype systems; and some flow into current systems assets through modernization programs improving their efficiency and effectiveness. The cost of current assets flows into net ownership costs which provide the resources to operate the fleet and shore establishment. The level of funds flowing into operations is a major determinant of the level of Naval readiness; that is, if less funds flow than are required, readiness is less.

Readiness resources needed can be defined as the amount of funds required to maintain operational commitments including adequate supply of technical spare parts, of personnel skills and grades for manning; of maintenance funds for ship and aircraft overhauls; of training in formal schools, tactics and fleet exercises.

Figure 6 is a more detailed representation of the R&D funds flow interactions in the cost of current assets area. With an increase or decrease in force assets, operational support costs increase or decrease, but not proportionally. The relative costs of maintenance, operating schedules, and manpower affect

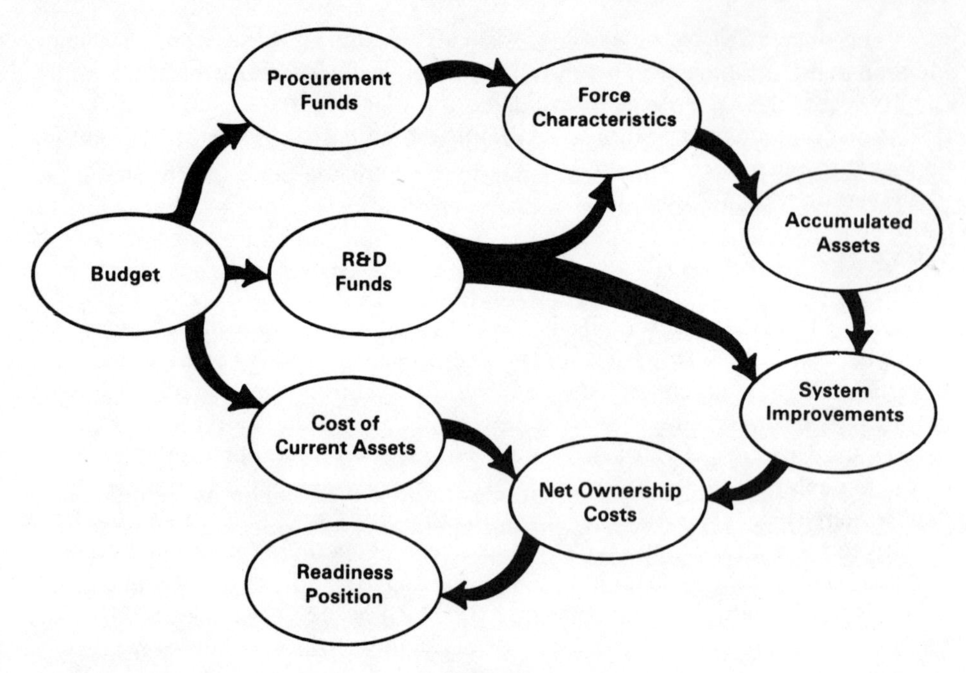


Figure 5

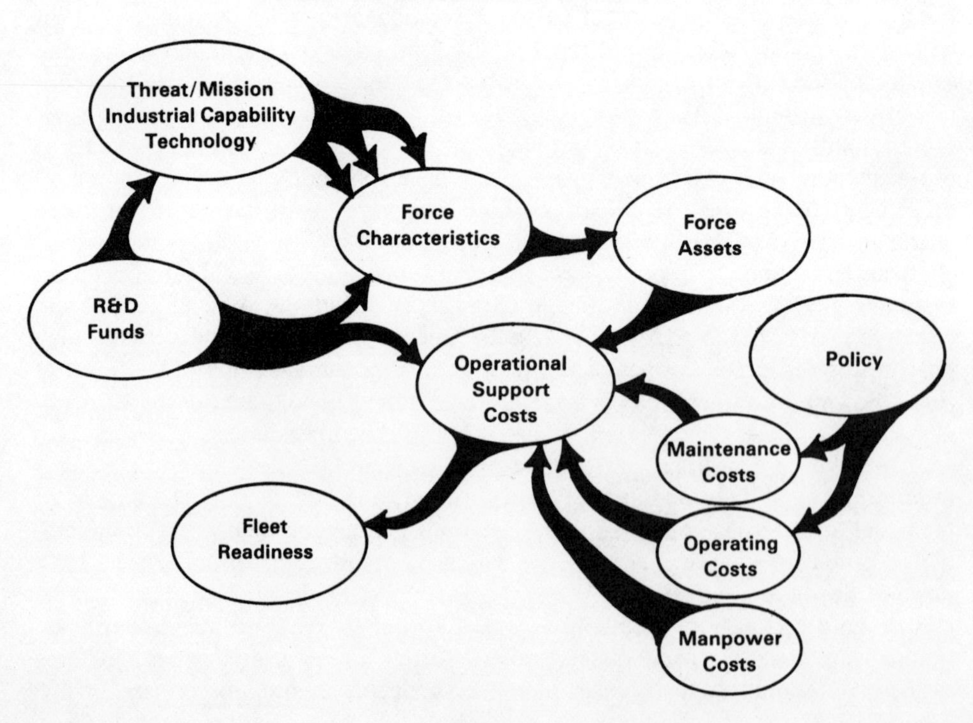


Figure 6

the degree of change in operation support costs as these new assets are acquired. In addition R&D funds, are expanded for modernization which should reduce the operational costs.

Each of these areas—maintenance, manpower and operations, are candidates for research in R&M. They are dependent on each other. Improvements in one can have either positive or negative effects on the others. Care must be taken to understand what the overall influence is on the entire weapon systems life. Serious investments to reduce the life-cycle cost of weapons can have a significant impact on the future composition of Naval forces. Given Figs. V and VI, consider the following scenario. As the Navy receives its budget, a determination is made to fund the current assets account at 100% of requirements; the remainder of the budget is apportioned at 20% for R&D and the remainder to procurement. If we translate this to the FY-81 Navy budget, we would have something on the order of 23 billion on current operations; 4 billion in R&D and 16 billion on procurement out of a 43 billion dollar total Navy budget. A further breakdown of current operations would include 16 billion for operations and maintenance and 7 billion for personnel. If over the next 5 years the Navy budget increases by 7% and by improving R&M related activities, the cost of current operations could be held to a 5% increase, the distribution for FY-86 would be 29 billion for current operations; 6 billion for R&D, and 23 billion for procurement, out of a 58 billion dollar total Navy budget. That two percent decrease in current operations' requirements costs increases the procurement account by more than 6 percent. Since current operations costs are such a large part of the budget, a small percentage gain can have a large influence on the remaining components of the budget. While the example shown indicated a decrease in current operations, it is also true that an increase in current operations costs has a major effect in reducing the ability to procure new assets. In fact, the latter case is our current difficulty, in that the funds required for current operations has been increasing at a rate that is seriously eroding the asset base.

## THE ACQUISITION CYCLE

If we want to interact with the acquisition cycle, where are we most effective? Figure 7 describes the relationship between product innovation and process innovation. Most will agree that product innovation is early in the cycle, that technology improvement spawns new products which then are considered as alternatives to system design. As you move towards advanced development where demonstrations, validations and cost estimating are completed, new product alternatives decrease quickly as full-scale development begins and process innovation peaks. There is considerable concern about process innovation or producibility. It seems that actual practice places the process innovation later in the cycle, which tends to lengthen the full-scale development period and causes process innovation to peak at the beginning of production. Another way of stating the concern is that process innovation is of little concern to the manufacturers until after full-scale development has been completed and DSARC III (the decision to move to full production) has been approved.

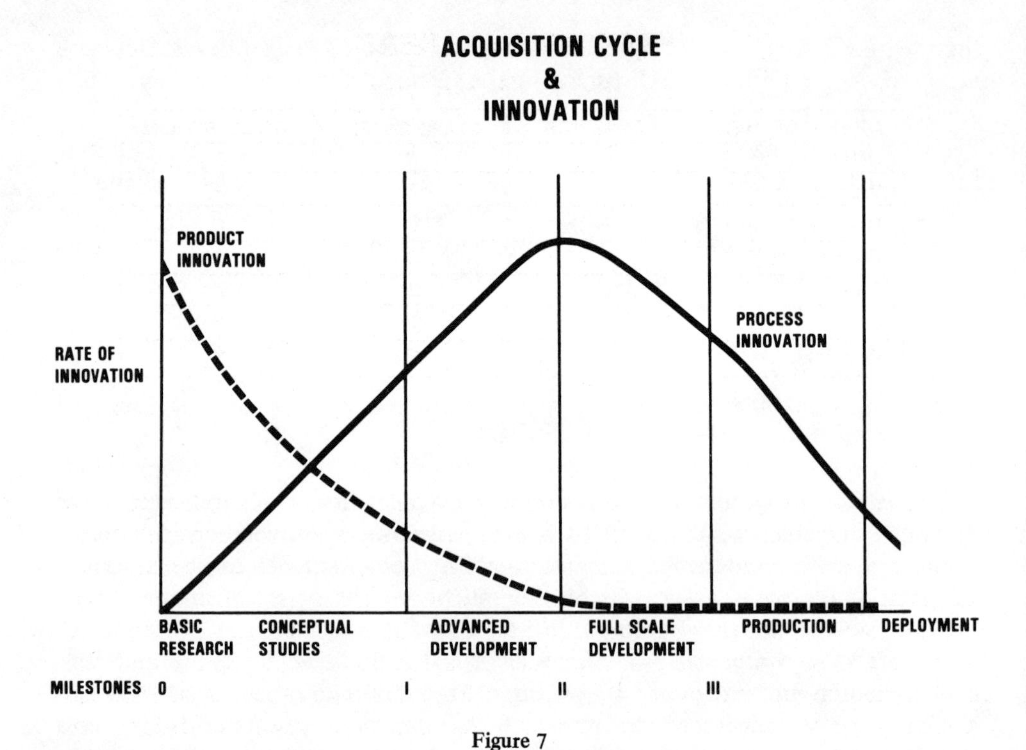


Figure 7

The ability to influence system characteristics must be accomplished early in the cycle. These new products are the results of both hardware as well as methodology development coming from the technology base. They are evaluated and integrated into the system design during the conceptual effort. While hardware development is important for meeting future threats, methodology development has a significant role in attempting to balance the conflicts and constraints into the best overall design and life-cycle operation of the system.

## R&M Areas of Need

Figure 8 displays several of these conflicts and constraints over three primary functional areas for system characteristics. The logistics area represents the life of the system and its ownership costs after procurement. Manpower is also a lifetime cost and is separated from the more traditional logistics considerations because of its increasing importance in manning systems and the demographics we discussed earlier. Acquisition represents those actions and activities that are required to bring the system into being. As one reflects on the three functional areas, it is obvious that they are not independent and actions in one area can influence the other areas.

The particular areas of need are an indication of the R&M activity which can influence each of the functional areas, and as a rough scale. A "P" relates to primary influence while an "S" suggests a secondary influence. Of those needs displayed, availability is the only one considered as having secondary

**FUNCTIONAL AREAS**

| AREAS OF NEED | LOGISTICS | MANPOWER | ACQUISITION |
|---|---|---|---|
| R&M | P | P | P |
| Warranties | P | P | P |
| Quality Control | P | P | P |
| Availability | P | S | S |
| T&E | S | S | P |
| L.C.C. | P | P | P |
| $P^3$ I | P | S | P |

Figure 8

influence for manpower and acquisition, yet availability is not independent of reliability and maintainability. This is especially true when you consider operational availability opposed to supply availability. Similar cases can be described for each of the other areas, which infer that there are many constraints placed on the "system designer" before the best affordable system can be produced. There are design to costs and life-cycle considerations, producibility and risk-cost tradeoffs and manpower constraints. Consider warranties: If new methodology can be developed to improve our ability to evaluate future cost avoidance due to increased reliability/availability early in the acquisition phase (before production), direct savings could be realized in the logistics and manpower areas during system operating life. Such methodology should contain concepts for rewarding both the Navy (through reduced life-cycle cost) and the manufacturers (through improved financial incentives during production). This would include a reduction in spare parts support; in organizational, intermediate and depot maintenance; and training and in support personnel at all levels. This is especially important for personnel, when you consider that it takes about three inductees to grow one E-6 electronics technician.

Reliability and maintainability considerations are important components in the total acquisition process. They and their related "ilities" that comprise the assurance sciences disciplines, are necessary to provide the capability for developing, testing, evaluating and determining the appropriate components and the organization of those components into affordable systems.

The increased complexity of weapon systems with their highly integrated circuits; their design for practical degradation; the need for fault detection and fault isolation schemes and applications for quality control, optimal inspection scheduling and nondestructive testing requires increased coordination between the Navy, academic and industry to understand the conflicts and constraints and develop producible and affordable systems for our National defense missions.

# Some Recent Work Related to Renewal Theory

Walter L. Smith

*University of North Carolina at Chapel Hill*

**Abstract**

This paper will describe new results in two research areas in which renewal theory is either the central issue or the tool of major importance. In the talk, because of time limitations, the survey will be necessarily somewhat limited, but full information will be available in the paper. The topics to be discussed are: (a) Transient Regenerative Processes; (b) Asymptotic Behavior of Cumulative Processes. The theory of regenerative processes has been useful in a wide range of applications but, as originally formulated, gives no guidance in the transient case, when the probability of a recurrence of the regeneration point is less than one; under (a) will be described amendments to the main theory, in which it will be shown that with suitably modified interpretation much of the standard theory (including the asymptotic normality results) can apply in the transient case. There are several practical areas of application for these results. Under (b) will be discussed work which shows that for quite general cumulative processes the cumulants are asymptotically linear functions of time; such results have long been known to hold for renewal processes, but the new formulae greatly extend the scope of these ideas and should find many applications to real problems.

## 1. REGENERATIVE PROCESSES

The purpose of this talk is to expound some of the main features of certain stochastic processes which have been useful in applications, and with the development of which the author has been closely associated. The first topic to be introduced: *regenerative processes*, is not to be confused with the elegant and powerful mathematical theory developed by J. F. C. Kingman in various papers, and expounded in his book (Kingman, [1972]). It is unfortunate that these two, admittedly somewhat overlapping but nonetheless, in major emphasis, quite distinct studies should bear the same name.

The regenerative process of the present note is intended to model a wide range of phenomena in the "real" world. It is based on the fundamental notion of a *tour.* We also have in the back of our mind a space X, usually a Euclidean one of finitely many dimensions, in which our measurements or observations occur. For example, if we are studying the waiting time (a continuous real variable) of a customer in some congestion process, then X need only be a space of one dimension; if we are simultaneously studying the waiting times of, say, five customers at five different but interacting service gates, then X needs to be a space of five dimensions.

Having said all this, let us now attempt to explain what we mean by a tour: it consists of:

(a) A *duration* $\mathscr{X} > 0$; $\mathscr{X}$ is a real-valued, positive, random variable, and successive values of $\mathscr{X}$ are independent and identically distributed.

(b) A function T(t), taking values in the space X, which we shall call the *graph*, of the real variable $t$ (for "time"), $0 \leqslant t < \mathscr{X}$.

(c) A *terminal value* $Y$ which is often, though not necessarily, some sort of limit of $T(t)$ as $t \uparrow \mathscr{X}$.

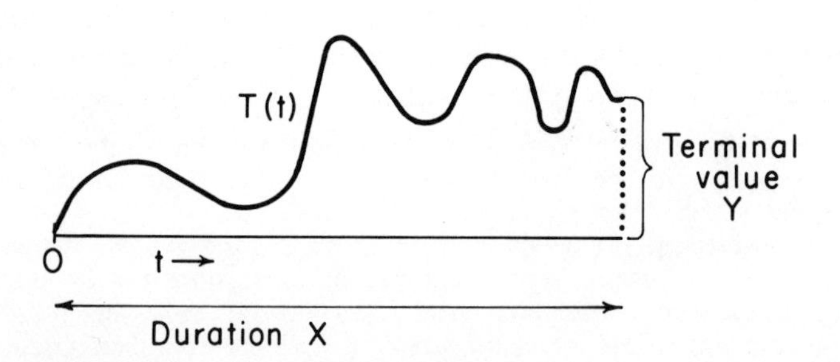


We shall write $B$ for a typical tour and $B_1, B_2$, ..., and so on, for a sequence of independent and identically distributed tours. There is very little mathematical difficulty in setting up a suitable measure-theoretic framework for these concepts, but in the present exposition it would seem both pretentious and off-putting.

In a picturesque way we can think of "Nature" as selecting a tour from a large "urn" of tours; this tour is then allowed to run its course for the period of time $X_1$ (the first "duration"). At the end of this first tour a second tour is instantly selected from the "urn" and its graph is abutted to the end of the first tour. It should be clear how it is imagined that the process develops, and that successive selections are supposed to be independent.

Thus we imagine the overall *regenerative process* to be a concatenation of tours. If we define $S_0 = 0$ and then write

$$S_n = X_1 + S_2 + \ldots + X_n, \text{ for } n \geqslant 1,$$

then we call the time instants $\{S_n\}$ *regeneration points* of the process. We shall write $S^*(t)$ for the "latest" regeneration point prior to (but including) the time-instant $t$, thus $S^*(t) = S_k$ if $S_k \leqslant t < S_{k+1}$. A picture of the one-dimensional regeneration process is then as follows:

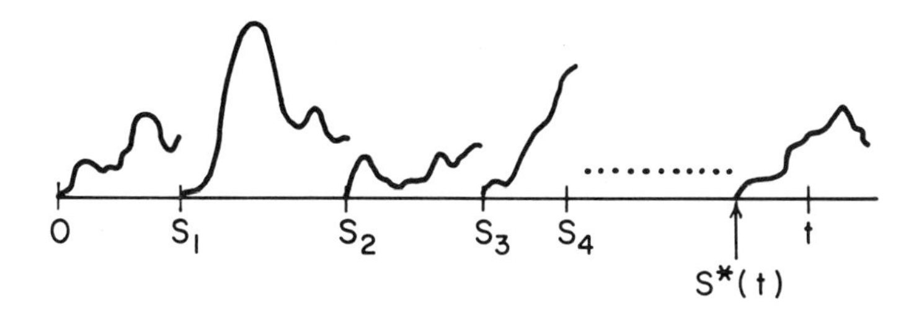


*Typically* $T(t)$ is nondecreasing and represents, for example, fuel consumption since the start of a tour, or the cost incurred since the start of the tour, or some other nondecreasing parameter associated with the process. However, $T(t)$ might also represent the number of customers needing service, or the noise level in some electronic monitoring system (for which the regeneration points would be, perhaps, the instants at which a new or newly serviced "black box" is installed.

## 2. CUMULATIVE PROCESSES

For simplicity of exposition let us assume that X is one-dimensional. Let us, copying an already established notation, write $T^*$ for the graph associated with the tour which is current at the instant "t." If we also write $N(t)$ for the number of tours completed by time $t$ then we can, for $t \geqslant X_1$, introduce a process:

$$W(t) = \textstyle\sum_{j \leqslant N(t)} Y_j + T^*(t - S^*(t)).$$

When $t < X_1$, we set $W(t) = T_1(t)$, where $T_1$ is the first graph.

This process, based on the underlying regenerative process, is called a *cumulative process.* Such processes are of considerable practical interest; the value $W(t)$ is the sum of all "terminal values" incurred, prior to (and including) the instant $t$, plus the current value of the graph function. For example, each terminal value $Y_j$ could represent the cost of the immediately preceding tour less salvage value of the equipment replaced at the end of the tour; the current value of the graph $T$ represents the cost incurred so far in the current tour (current, that is, at the instant $t$). Thus, in this example, $W(t)$ would be the total cost incurred up to the time instant $t$.

To explain our results further we shall make need to introduce certain notations:

$$F(x) = P(X_j \leqslant x), \quad x \geqslant 0; \tag{2.1}$$

$$G(x,y) = P(X_j \leqslant x \ \& \ Y_j \leqslant y), \quad x \geqslant 0, \ y \geqslant 0; \tag{2.2}$$

$$\alpha_r = E(X_j)^r; \tag{2.3}$$

$$s_r = E(U_j)^r; \tag{2.4}$$

$$\alpha_{rs} = E(X_j)^r(Y_j)^s; \tag{2.5}$$

$$\sigma_1^2 = Var\ X_j = \alpha_2 - (\alpha_1)^2; \tag{2.6}$$

$$\sigma_2^2 = Var\ Y_j = s_2 - (s_1)^2 ; \tag{2.7}$$

$$\rho\sigma_1\sigma_2 = Cov(X_j, Y_j) ; \tag{2.8}$$

$$F(s) = E \exp(-sX_j) ; \tag{2.9}$$

$$G(s,\theta) = E \exp(-sX_j + i\theta Y_j) ; \tag{2.10}$$

$$H(\theta) = E\exp(i\theta Y_j) . \tag{2.11}$$

We can now describe some long-established results concerning these cumulative processes. We will not burden the reader in the present account with all the technical conditions involved (which are mainly that any moments which appear in formulae must be finite); those who wish to obtain the rigorous details should read the original memoir on this subject (Smith [1955]).

**Result A**

As $t \to \infty$, with probability one,

$$W(t)/t \to s_1/\alpha_1,$$

This result is really part of an *ergodic* theorem for cumulative processes; the theorem is saying rather more than that if we observe a process for a "very long" time period "t" then, with high probability, $W(t)/t$ should be near the constant $s_1/\alpha_1$,

**Result B**

As $t \to \infty$

$$EW(t)/t \to s_1/\alpha_1,$$

This is the complementary part of the ergodic theorem; the two parts together say that, as $t \to \infty$,

$$[EW(t)/t] - [W(t)/t] \to 0.$$

Thus, in the physicist's terms, temporal and phase averages can be made arbitrarily close by taking $t$ large enough.

**Result C**

As $t \to \infty$,

$$\text{Var } W(t) \sim (\gamma t)\alpha_1,$$

where

$$\begin{aligned}\gamma &= \sigma_2^2 - 2\rho\sigma_1\sigma_2(s_1/\alpha_1) + \sigma_1^2 (s_1/\alpha_1)^2,\\ &= \text{Var } [Y_1 - (s_1/\alpha_1)X_1].\end{aligned}$$

This relatively simple asymptotic formula for the variance of the cumulative process can be useful in making rough-and-ready judgments about the sort of fluctuations to be expected of $W(t)$ if the model were a valid representation of the true process; there are also theoretical uses for the formula.

**Result D**

If we employ the notation already established, then, as $t \rightarrow \infty$ ,

$$[W(t) - [s_1 t / \alpha_a] / [\gamma t / \alpha_1]^{1/2}$$

is asymptotically N(0,1). This is the Central Limit Theorem for cumulative processes, and is useful in making inferences, albeit approximate ones, about the process $W(t)$. Note that results *C* and *D* harmonize over the value to be used for Var $W(t)$, but *C* requires separate proof and cannot be deduced from *D*.

There are multivariate extensions of the two results *C* and *D*. We indicate here merely an intuitive way in which these generalizations can be proved. Let $Y$ now stand for the k-dimensional vector

$$(Y^{(1)}, Y^{(2)}, \ldots, Y^{(k)}),$$

and suppose we now consider the *vector*-valed cumulative process

$$W(t) = \textstyle\sum_{j < N(t)} Y_j + T^*(t - S^*(t)),$$

where $T$ is now a *vector*-valued graph function.

If $(a_1, a_2, \ldots, a_k)$ be a "dummy" vector with real coordinates we can consider the one-dimensional cumulative process

$$a_1 W^{(1)} + a_2 W^{(2)} + \ldots + a_k W^{(k)}$$

and apply the results *C* and *D* to this real-valued process.

Evidently we need calculate whatever corresponds to the constant $\gamma$ in the earlier results. To do this let us set

$$u_j = \text{Cov } (Y_1^{(j)}, X_1)$$
$$v_{ij} = \text{Cov } (Y_1^{(i)}, Y_1^{(j)});$$

Then, in an obvious extension of our notation, we also write

$$s_1^{(j)} = EY_1^{(j)},$$

In terms of these constants we find that

$$\gamma = \sum_i \sum_j a_1 a_j \gamma_{ij},$$

where, it turns out,

$$\gamma_{ij} = \nu_{ij} - (u_i s_1^{(i)} + u_j s_1^{(j)})/\alpha_1 + \sigma_1^2 s_1^{(i)} s_1^{(j)}/\alpha_1^2.$$

Thus we can make the following inferences (which can be rigorously proved):

**Result C***

As $t \to \infty$,

$$\mathrm{Cov}[W^{(i)}(t), W^{(j)}(t)] \sim \gamma_{ij} t/\alpha_{ij} ;$$

**Result D***

As $t \to \infty$ , the quantities

$$[W^{(j)}(t) - s_1^{(j)} t/\alpha_1]/[t/\alpha_1]^{1/2},$$

for $j = 1, 2, \ldots, k$, are asymptotically distributed in a multivariate normal distribution, with zero means, and covariance matrix $[\gamma_{ij}]$.

## 3. MORE DETAILED RESULTS

Let us return to the question of the real-valued cumulative process. Much current research effort of the author has been devoted to seeing how much more can be said about the asymptotic behavior of $W(t)$; the desire has been to obtain rather more "fine-structured" results than the results *B*, *C*, and *D*, which give, as it were, only the "dominant" behavior.

How much more can be said about $W(t)$ depends on what moments $\alpha_{rs}$ can be assumed to exist. We shall focus our attention on the cumulants of $W(t)$; the first few of these inform us about the mean, the variance, the skewness, and the kurtosis, of the distribution of $W(t)$. In the event that we can derive reasonably accurate formulae for these cumulants then, for instance, the Edgeworth expansion (or one of its proprietary rivals) could be used to provide more precise information about the asymptotic distribution of $W(t)$ than that given by the bald Central Limit Theorem (Result *D* above). Indeed, the way may be open to tackle inferential problems involving cumulative processes.

Moments, rather than cumulants, of $W(t)$, may have more immediate, more direct, appeal. However, it turns out that very simple asymptotic formulae hold for the cumulants of $W(t)$, formulae that are much more convenient to deal with than the corresponding ones would be for the moments.

A critical role is played in this part of the study by

$$\mathbf{T}(s,\theta) = \mathbf{E}\int_0^{X_1} \exp\{-st + i\theta T_1(t)\}\,dt,$$

where $T_1(t)$ is the initial, but typical, graph and $X_1$ the associated duration. This function gives rise to some curious moment-like numbers:

$$\rho_{jk} = \mathbf{E} \int_0^{X_1} t^j [T_1(t)]^k dt.$$

Presumably the assumptions one makes about the underlying model will allow the calculation, at least in theory, of these numbers $\{\rho_{ij}\}$.

Ideally one would like to know the characteristic function $\Phi(\theta,t)$ of the process $W(t)$, given by

$$\mathbf{E} \exp [i\theta W(t)].$$

There seems to be no straightforward formula for this quantity, but there does exist an elegant formula for its ordinary Laplace Transform:

$$\Phi^0(\theta,s) = \frac{\mathbf{T}(s,\theta)}{1 - \mathbf{G}(s,\theta)}$$

Let us write $K_n(t)$ for the $n^{\text{th}}$ cumulant of $W(t)$, then, as $t \to \infty$,

$$K_n(t) = A_n t + B_n + r_n(t),$$

where $r_n(t) \to 0$ as $t \to \infty$, and $A_n$ and $B_n$ are multinominals in the product-moments

$$\{\rho_{(j-1),k}\} \text{ and } \{\alpha_{j,k}\},$$
$$\text{for } 1 \leqslant j \leqslant n + 1,$$
$$0 \leqslant k \leqslant n + 1 - j.$$

The numbers $A_n$ depend only on the numbers $\{\alpha_{rs}\}$ and not at all on the numbers $\{\rho_{rs}\}$; this can be seen since the numbers $A_n$ plainly help describe the dominant behavior of $W(t)$ and the numbers $\{\rho_{rs}\}$ are concerned with second-order behavior (and therefore *do* appear in the constants $B_n$).

The crucial theoretical problem in the present context, and one which calls for quite elaborate Fourier analysis, is to discover the way in which the speed at which $r_n(t)$ tends to zero as $t \to \infty$ is related to the smallness of the tails of the distributions of $F$ and $G$. On the other hand, form the point of view of applications, the prime interest focuses on the calculation of the constants $A_n$ and $B_n$. This is tedious for large values of $n$ although for $n \leqslant 4$ the formulae are not too off-putting. The computations are made easier by arranging the time scale to make $\alpha_1$ equal unity. One then introduces the new random variables

$$V_n = Y_n - s_1 X_n,$$

and calculates the product moments:

$$\lambda_{rs} = \mathbf{E}(X_n)^r (V_n)^s.$$

It is then possible to develop equations connecting, by means of these new product-moments, each $A_n$ with the $\{A_k\}$ for all $k < n$. We will not give details of these computations in this talk, referring the reader to the technical report, which is in preparation and will soon be available, for details (Smith [1981], see also Smith [1979]). However, to give some idea of these equations we quote one or two. The statement of these equations is somewhat simplified if it is assumed that the scale on which "V" is measured is chosen to make $\lambda_{02} = 1$ (it should be clear that $\lambda_{01} = 0$). One then finds that $A_1 = A_2 = 1$, and then that

$$A_3 = \lambda_{03} - 3\lambda_{11}$$

$$A_4 = \lambda_{04} - 6\lambda_{12} - 4A_3\lambda_{11} + 3\lambda_{20}.$$

To provide some idea of how involved these formula become as $n$ increases we quote the equation for $A_8$:

$$\begin{aligned} A_8 = \lambda_{08} &- 28\lambda_{16} + 210\lambda_{24} - 420\lambda_{32} \\ &+ 105\lambda_{40} - 56\lambda_{15}A_3 + 560\lambda_{23}A_3 - 840\lambda_{31}A_3 \\ &+ 280\lambda_{22}(A_3)^2 - 200\lambda_{30}(A_3)^2 - 70\lambda_{14}A_4 \\ &+ 420\lambda_{22}A_4 - 210\lambda_{30}A_4 + 280\lambda_{21}A_3A_4 + 35\lambda_{20}(A_4)^2 \\ &- 56\lambda_{13}A_5 + 168\lambda_{21}A_5 + 56\lambda_{20}A_3A_5 - 28\lambda_{12}A_6 + 28\lambda_{20}A_6 \\ &- 8\lambda_{11}A_7. \end{aligned}$$

There are 21 terms on the right-hand side of this equation; the corresponding one for $A_{10}$ has 41 terms on the right, and the parallel equations for the constants $B_n$ are much worse than those for the $A_n$!

Before the days when computers, and especially microcomputers, were easily available, formulae such as we are now contemplating would have little value in applied work; they would be tedious to calculate and a fruitful source of numerical slips. The matter is different now. Indeed, after tediously computing the equations for the $A_n$ and $B_n$ in the range $n \leqslant 8$, the author wrote a program in Basic which computed the actual algebraic equations for $n \leqslant 10$, and *printed* them in a ready-to-use form. It should be possible to write a program which will not only do this calculation of algebraic formulae but will itself *write a program* to compute numerical values for the $A_n$ and $B_n$. Once this stage has been reached the actual equations, such as the one immediately above, will become mere curiosities, and the calculation of these asymptotic cumulants will seem in no way off-putting. It will then become an easy task to study the effects of changing parameters of the underlying process.

As we have implied, the formulae for the lower order cumulants are *not* horrendous; the one for the asymptotic variance of $N(t)$, the renewal count, is of particular interest. If $\alpha_3 < \infty$, then as $t \to \infty$,

$$\text{Var } N(t) = (\sigma^2/\alpha_1^2)t + (5\alpha_2^2/4\alpha_1^4) - (2\alpha_3/3\alpha_1^3) - (\alpha_2/2\alpha_1^2) + o(1).$$

The formula for the general cumulative process $W(t)$ is not much more complicated.

A problem which has not yet been tackled, but which is plainly tractable, is the conversion of the results we have been discussing into results for the multivariate case. This would enable one to discuss with fair precision the variances and covariances, for instance, of simultaneous measurements (like cost and fuel consumption).

## 4. TRANSIENT REGENERATIVE PROCESSES

Suppose that in the regenerative model of the earlier sections we had $F(0+) < 1$. Another way of looking at this is to suppose that with a positive probability we could have $X_n = \infty$. This we shall call the case of the *transient* regenerative process. If $\epsilon$ is the regenerative event, then we now have the case when $\epsilon$ is not certain to recur.

Consider the following examples of situations in which transient regenerative events are the appropriate ones to consider:

1. We might be watching a renewal process, waiting for a long gap to occur (that is: waiting for a "lifetime" to exceed some prescribed constant $L$). Such a problem has been investigated by Lamperti [1961].
2. In the study of queuing times, when the traffic intensity is less than one (the *stable* case) the stationary distribution of queuing time can be shown to be the sum of a geometrically distributed number of "ladder" variables. This is equivalent to studying the distribution of the time to the last occurrence of $\epsilon$ in a transient regenerative process.
3. Consider a simple model for telephone traffic in which demands for calls come in the form of a Poisson process, while the durations of conversations (assumed independently distributed, one from the others) have some fixed arbitrary distribution. A little reflection will show that the time to the first "lost" call (one which falls in the period of a conversation started earlier) presents the same problem as encountered in the previous paragraph.
4. If we study the Type II Geiger Counter in which the arrivals of particles cause prolongations of the instrument's dead period, and ask about the distribution of the interval between successively *recorded* particles, then it transpires that we are studying once again the problem of the previous paragraph. This problem is given some discussion in Feller [1971].

We shall give a sketch of part of the theory for these transient processes. Indeed, in this talk we shall only discuss in any detail the case when there exists a $\sigma > 0$ such that

$$\mathbf{E} \exp \sigma X_n = 1.$$

This case, which shall call the "good" case, is the one of major interest, and the state of the theory when such a $\sigma$ does not exist is not wholly satisfactory. However, in the good case, it will be seen that if we define a probability measure $\tilde{F}$ by the requirement $e^{\sigma x} dF = d\tilde{F}$ then $\tilde{F}$ is *proper*. This device, which is hardly new, enables one to replace the study of the transient cumulative process by that of a familiar "proper" one.

We have already defined $S^*(t)$ as the time of the latest regeneration point of the process prior to (or including) the instant $t$, let us also write $S^\dagger(t)$ for the first regeneration point to occur strictly *after t.* Then $S^\dagger(t) - t$ we shall call the "forward delay" and write it $\tau(t)$. It is well known that when the first moment $\alpha_1$ is finite then the distribution of $\tau(t)$ tends to a limiting form; we shall write $K(x)$ for this limit distribution function.

We can say the original transient process is "alive" at time $t$ if the $S^\dagger(t)$ is finite; the, perhaps illogical, idea being that when it is recognized at a regeneration point that the lifetime about to start will be of infinite length we shall say the process has "died." We shall now introduce the stochastic process $\mathbf{S}(t)$ by the requirement that

$$\begin{aligned} \mathbf{S}(t) &= 1 \text{ if the process is ``alive'' at time } t, \\ &= 0 \text{ if the process is ``dead'' at time } t. \end{aligned}$$

Another process needs to be introduced, and it really requires a somewhat elaborate measure-theoretic set-up to be defined rigorously. We shall here content ourselves with a heuristic description. We shall say the process $\mathbf{G}(t)$ is *natural* if it depends only on the tours that have been sampled up to and including the instant $t$, and note that this includes the tour current at time $t$.

By an obvious extension of our notation $\tilde{\mathbf{E}}$ will mean an expectation based on the probability measure derived from $\tilde{F}$. It is then possible to show that

$$\mathbf{E}\{\mathbf{G}(t)|\mathbf{S}(t) = 1\} = \frac{\tilde{\mathbf{E}}e^{-\sigma\tau(t)}\mathbf{G}(t)}{\tilde{\mathbf{E}}e^{-\sigma\tau(t)}}.$$

Let $\chi(A)$ be the usual indicator function of the event $A$. Then if $A(t)$ be an event depending on the time $t$ we shall say it is a *sluggish* event if, for every fixed $T$, as $t \pm \infty$,

$$\mathbf{P}\{\chi(A(t)) \neq \chi(A(t+T))\} \to 0$$

Following on this idea we shall say a process $\mathbf{G}(t)$ is sluggish if all events like $\{\mathbf{G}(t) \leqslant x\}$, for every fixed real $x$, are sluggish.

These ideas are very useful in studying the transient cumulative process. We shall end this talk with one example, the most important one. This concerns the cumulative process $W(t)$ which is such that $\mathbf{E}W(t) = 0$ for all $t$, with all relevant second moments finite. Then it is possible to show that $W(t)/\sqrt{t}$ is both sluggish and natural. But if a process is both sluggish and natural one can show that

$$\frac{\tilde{\mathbf{E}}e^{-\sigma\tau(t)}\mathbf{G}(t)}{\tilde{\mathbf{E}}e^{-\sigma\tau(t)}} - \tilde{\mathbf{E}}\mathbf{G}(t) \to 0.$$

Evidently this implies that

$$\mathbf{E}\{\mathbf{G}(t)|\mathbf{S}(t) = 1\} - \tilde{\mathbf{E}}\mathbf{G}(t) \to 0.$$

Thus it becomes a straightforward chore to translate results about the proper (recurrent) cumulative process into corresponding results about the transient one *conditional upon the latter process being still alive.* One can obtain in this way central limit theorems (i.e., conditional asymptotic normality) together with theorems about the conditional cumulants of the transient process. The full details of this work are available in the Ph.D. thesis of E. Murphree (directed by the present author) (Murphree [1981]; it is anticipated that various parts of this thesis will ultimately be published.

By way of illustration: we have seen that $W(t)/\sqrt{t}$ is sluggish and natural. Consider the event, for some arbitrary fixed $x$,

$$A(t) = \{[W(t)/\sqrt{t}] \leqslant x\}.$$

Then $\mathbf{G}(t) = \chi(A(t))$ is also a sluggish and natural process. Hence we can deduce that

$$\mathbf{P}\{A(t)|\mathbf{S}(t) = 1\} \rightarrow \mathbf{\Phi}(x),$$

because $\tilde{\mathbf{P}}\{A(t)\} \rightarrow \mathbf{\Phi}(x)$, where $\mathbf{\Phi}(x)$ is the usual standard normal distribution function. Thus, conditionally upon the process being alive at time $t$, $W(t)/\sqrt{t}$ is asymptotically $\mathbf{N}(0, 1)$.

When we do not have the "good" case (no suitable $\sigma$ existing) we have what is called the sub-exponential case. Here the story is much less satisfactory. One has to assume the distribution $F$ belongs to the Chistyakov class (see Chistyakov [1964]). To belong to this class, if $F_2$ represent the Stieltjes convolution $F*F$ then the following must hold as $x \rightarrow \infty$:

$$2[1 - F(x)] \sim 1 - F_2(x).$$

One surprising difference between the "good" case and the sub-exponential case is that in the latter case moments of $N(t)$, even when conditioned upon the process being alive at time $t$, converge to finite limits as $t \pm \infty$. Indeed, what happens in the sub-exponential case is dramatically different from what happens in the "good" case.

## REFERENCES

Chistyakov, V.P. (1964), A theorem on sums of independent positive random variables, *Theory of Probability and its Applications*, 9, 640-648.

Feller, W. (1971), *An Introduction to Probability Theory and Its Applications, Volume 2, Second Edition*, New York: Wiley.

Kingman, J.F.C. (1972), *Regenerative Phenomena*, New York: Wiley.

Lamperti, J. (1961), A contribution to renewal theory, *Proceedings of the American Mathematical Society*, *12*, 724-731.

Murphree, E.S. (1981), *Transient Cumulative Processes*, Ph.D. dissertation: University of North Carolina at Chapel Hill.

Smith, W.L. (1955), Regenerative stochastic processes, *Proceedings of the Royal Society, Series A*, *232*, 6-31.

Smith, W.L. (1979), On the cumulants of cumulative processes, Institute of Statistics Mimeo Series No. 830, University of North Carolina at Chapel Hill.

Smith, W.L. (1981), Asymptotic behavior of cumulative processes, to appear shortly as technical report.

# Part II: Systems Reliability

# Approximate Interval Estimates for System Reliability Using Asymptotic Expansions

Alan Winterbottom

*City University, London*

**Abstract**

For systems with general structure functions, possibly with repeated components, the construction of interval estimates for system reliability, based on component test data, can cause considerable computational difficulties even when the structure is relatively simple. For the classical approach, the asymptotic normality of large sample estimators has generality, but is not sufficiently accurate for sample sizes of practical interest. Improvements in accuracy can be effected by developing asymptotic expansions which provide successive corrections to the large sample formula.

The Bayesian approach to the problem is of growing interest, but it has not received the attention accorded the classical approach. For one case, that of a series system, the posterior distribution of system reliability can be obtained by inverting a Mellin integral transform. However, this technique also incurs computational difficulties as the number of components increases and it is not available for more complicated structures. For this special case it is shown that Cornish and Fisher expansions for percentiles can be developed and lead to very accurate approximations even for small numbers of components.

For general structures an adaption of the classical asymptotic expansion method is proposed. Whereas the asymptotic variable in the Cornish and Fisher formulation for series systems is $m$, the number of components, the asymtotic variable for general structures is $n = \min n_i$. Since series structures are in the class of monotonic structures considered it will be of interest to compare the two expansions, the one in $m$ and the other in $n$.

## 1. INTRODUCTION

Severe computational difficulties often arise when exact procedures are used to construct interval estimates for system reliability based on component test data.

Even for the few relatively straightforward cases, such as series systems with binomial or exponential test data, calculations become very difficult as the number of components and as test sample sizes increase. These difficulties are largely due to the nuisance parameter aspect of the problem which arises because the test data provide information directly on the component reliabilities (the nuisance parameters) and only indirectly on the system reliability (the parameter of interest) through the structure function of the system.

Many methods have been proposed with the objective of providing good approximate intervals together with ease of computation. Most of these can only be used for a restricted range of cases. Two large sample methods are available for general use namely the asymptotic normality of the maximum likelihood estimator and the asymptotic chi-squared distribution of the likelihood ratio test statistic. The latter gives the better approximations but they are not satisfactory for sample sizes of practical interest. However, intervals based on large sample normality can be effectively improved by asymptotic expansions which provide successive adjustments (corrections) to the large sample formula. In Section 2 the methodology is described through its application to classical confidence intervals. Explicit expansions are given for two important cases and the results of a simulation study are included for two structures for which exact procedures are intractable.

In Section 3 an account is given of the way in which the expansions could be adapted to provide good approximations for the percentage points of posterior distributions of system reliability, enabling improved approximate Bayesian intervals to be determined. For a series system, it is possible to apply, more or less directly, the closely related Cornish and Fisher asymptotic expansions to improve approximate Bayesian intervals. This application is given in detail and the numerical results show very accurate approximations even for small numbers of components and small test sample sizes.

## 2. METHODOLOGY FOR THE EXPANSIONS AND SOME APPLICATIONS

Let $\theta_i$ be the reliability of the $i$th component and $\theta = \psi(\theta_1, \ldots, \theta_m)$ be the reliability of the system where $\psi$, the system reliability function, is monotonic increasing in each $\theta_i$. Suppose $T_i$ is an estimator of $\theta_i$ such that $T_i - \theta_i$ has first cumulant of order $n_i^{-1}$ and rth cumulant of order $n_i^{1-r} (r \geqslant 2)$. The $n_i$ are sample sizes or possibly numbers of failures and we note here that it is sometimes more convenient to take the $\theta_i$ to be parameters related to reliability such as mean times to failure. The cumulant property is that required for the development of the expansions of Cornish & Fisher (1937).

Consider $T = \psi(T_1, \ldots, T_m)$ as an estimator of $\theta = \phi(\theta_1, \ldots, \theta_m)$ and let $n = \min(n_i)$ so that $n_i = n\lambda_i$, $\lambda_i \geqslant 1 (i-1, \ldots, m)$. As $n$ increases indefinitely, with the $\lambda_i$ fixed, the limiting distribution of $n^{1/2}(T-\theta)/s$, where $s^2 = \sum_i v_i (\partial \psi / \partial \theta_i)^2 / \lambda_i$ and $v_i = \lim \operatorname{var}(n_i^{1/2} T_i)$ is standard normal. To this order of approximation the parameters in $s^2$ can be replaced by their corresponding estimators to give a random variable $S^2$ leading to large sample confidence limits given by $\tilde{\theta}(\xi) = T - \xi S/\sqrt{n}$, where $\xi$ is an appropriate standard normal value.

**Table 1 Cumulants of Z**

| 1 | $n^{-1/2}$ | $n^{-1}$ | $n^{-3/2}$ |
|---|---|---|---|
| | $\kappa_{11}$ | | $\kappa_{12}$ |
| $\kappa_{21}$ | | $\kappa_{22}$ | |
| | $\kappa_{32}$ | | $\kappa_{33}$ |
| | | $\kappa_{43}$ | |
| | | | $\kappa_{54}$ |

The first- and second-order adjustments are derived by considering the random variable

$$Z = n^{1/2} Y = n^{1/2}\{G_1/n + (H + G_2/n)(T-\theta) + G_3(T-\theta)^2 + G_4(T-\theta)^3\} \quad (2.1)$$

Here $H, G_1, \ldots, G_4$ are functions of $T_1, \ldots, T_m$ and of $\lambda_1, \ldots, \lambda_m$ only. From results due to James (1955) and James & Mayne (1962) concerning the cumulants of transformed random variables, both $T - \theta$ and $Y$ have the cumulant property in $n$. Table 1 exhibits cumulant coefficients beneath their associated powers of $n^{-1/2}$ in the cumulant expansions of $Z$ so that for example, the second cumulant is $X_2 = \kappa_{21} + \kappa_{22}/n + O(n^{-2})$.

The construction of (2.1) enables two adjustments to be made to the crude normal approximation. Setting $\kappa_{21} = 1$ determines $H$ and leads to the crude approximation $n^{1/2}H(T-\theta)$. The first adjustment is given by $\kappa_{11} = \kappa_{32} = 0$ from which $G_1$ and $G_3$ are obtained. The first-order adjusted approximation is then $n^{1/2}\{G_1/n + H(T-\theta) + G_3(T-\theta)^2\}$. Finally $\kappa_{22} = \kappa_{43} = 0$ yields $G_2$ and $G_4$ and the second-order adjusted approximation is (2.1) in full. The cumulant generating function of $Z$ is now $-t^2/2 + 0(n^{-3/2})$. Further adjustments may be obtained by suitably augmenting (2.1). The reversion to give asymptotic expansions for confidence limits is obtained by substituting $\theta = T + \tau_1/n^{1/2} + \tau_2/n + \tau_3/n^{3/2}$ in $Z = \xi$, where $\xi$ is a standard normal value. The $\tau_i$ are identified by equating to zero the coefficients of powers of $n^{-1/2}$ up to $n^{-3/2}$. Thus

$$\tau_1 = -\xi/H, \tau_2 = (G_3\xi^2 + G_1H^2)/H^3, \quad \tau_3 = \{(G_4H - 2G_3^2)\xi^3 + (G_2H^3 - 2G_1G_3H^2)\xi\}/H^5 . \quad (2.2)$$

For binomial and exponential test data the moments are of the form

$$E(T_i-\theta_i) = 0, E(T_i-\theta_i)^2 = \mathrm{v}_i/n_i, E(T_i-\theta_i)^3 = w_i/n_i^2,$$
$$E(T_i-\theta_i)^4 = 3\mathrm{v}_i/n_i^2 + z_i/n_i^3 ,$$
$$E(T_i-\theta_i)^5 = 10\mathrm{v}_i w_i/n_i^3 + 0\,(n_i^{-4}),\ E(T_i-\theta_i)^6 = 15\mathrm{v}_i^3/n_i^3 + 0\,(n_i^{-4}) \quad (2.3)$$

For binomial data $\mathrm{v}_i = \theta_i\,(1-\theta_i)$, $w_i = \theta_i(1-\theta_i)\,(1-2\theta_i$ and $z_i = \theta_i(1-\theta_i)(1-6\theta_i + 6\theta_i^2)$, where $\theta_i$ is the proportion reliable in the population and $T_i$ is the sample proportion reliable. When failure times are exponential take $\theta_i$ to be the mean time to failure and $T_i$ to be the ratio of total time on test to $n_i$, where $N_i \geqslant n_i$ components of type $i$ are placed on test at time $t = 0$ and testing terminated at the $n_i$th failure. Then $v_i = \theta_i^2, w_i = 2\theta_i^3$ and $z_i = 6\theta_i^4$. No higher moments are required in the derivation of first- and second-order adjustments. The above estimators are maximum likelihood estimators and have the required cumulant property. Further, their unbiasedness considerably simplifies the algebra.

General expressions for $H$, $G_1, \ldots, G_4$ and hence for $\tau_1\tau_2$ and $\tau_3$ have been given by Winterbottom (1980). For a series system with binomial component test data we have $\theta = \Pi_i\theta_i$ and $T = \Pi_i T_i$, where $T_i$ is the proportion of reliable components in the $n_i$ tested components of type $i$. Thus $v_i = \theta_i(1-\theta_i)$ and $w_i = \theta_i(1-\theta_i)\ (1-2\theta_i)$. For discrete test data an exact procedure yields lower limits satisfying the probabilistic inequality Prob $(\tilde{\theta} < \theta) \geqslant \gamma$, where $\gamma$ is the confidence level. When sample sizes are small the confidence probability may substantially exceed the confidence level for particular values of the parameter vector. There is therefore little point in deriving a second-order adjustment since the accuracy of approximation may be diminished.

Denoting $\sum_i (1-T_i)/(\lambda_i T_i)$ by $\sigma^2$ the large sample formula for confidence limits is

$$\tilde{\theta}(\xi) = T(1-\xi\sigma/n^{1/2}) \tag{2.4}$$

and the first order adjusted formula is

$$\tilde{\theta}(\xi) \simeq T(1-\xi\sigma/n^{1/2} - [(\xi^2+1)\{\sum_i(1-T_i^2)/(\lambda_i^2 T_i^2) - 3\sigma^4\} + \sum_i(1-T_i)1-2T_i)/(\lambda_i^2 T_i^2)](6n\sigma^2)^{-1} . \tag{2.5}$$

Table 2 demonstrates the improvement gained by using (2.5) rather than (2.4). Results obtained using the likelihood ratio method are also shown. Lower limits using an exact method have been obtained from the tables of Lipow and Riley (1960).

**Table 2 Lower 90% confidence limits for series systems and binomial testing**

| n | Failures | Likelihood ratio | (2.4) | (2.5) | Tables |
|---|---|---|---|---|---|
| 10 | 1 1 | 0.629 | 0.655 | 0.616 | 0.607 |
| | 2 2 | 0.451 | 0.457 | 0.443 | 0.445 |
| 20 | 2 2 | 0.687 | 0.701 | 0.681 | 0.683 |
| | 1 2 2 | 0.643 | 0.654 | 0.638 | 0.639 |
| | 1 2 3 | 0.596 | 0.605 | 0.593 | 0.595 |
| 50 | 1 1 2 | 0.865 | 0.874 | 0.862 | 0.862 |
| | 1 2 4 | 0.798 | 0.805 | 0.795 | 0.789 |
| 100 | 1 1 2 | 0.931 | 0.936 | 0.929 | 0.929 |
| | 2 3 5 | 0.861 | 0.866 | 0.860 | 0.858 |

When considering exponential times to failure the numbers of failures are usually small compared to the sample sizes used in binomial testing. In order to gain effective improvements to the large sample approximation it is necessary to derive two adjustments. Consider the series system with exponentially distributed times to failure. Let $\theta = \Sigma_i 1/\theta_i$, where $\theta_i$ is the mean time to failure of the *i*th component. Then $\theta$ is the system failure rate and the system reliability for mission time $t_m$ is $e^{-\theta t_m}$. Also $v_i = \theta_i^2, w_i = 2\theta_i^3$ and $z_i = 6\theta_i^4$. The asymptotic expansion yielding approximate confidence limits for $\theta$ is

$$\begin{aligned}\tilde{\theta}(\xi) &= \Sigma T_i^{-1} - \xi\{(\Sigma\lambda_i^{-1}\, T_i^{-2})\,/\,n\}^{1/2} \\ &\quad + \{(\xi^2+2)\ (\Sigma\lambda_i^{-2}\, T_i^{-3})\,/\,(3\Sigma\lambda_i^{-1} T_i^{-1}\}n^{-1} \\ &\quad - [2(4\xi^3+17\xi)\ (\Sigma\lambda_i^{-2}\, T_i^{-3})^2/\ (\Sigma\lambda_i^{-1} T_i^{-2})^{5/2} \\ &\quad - 9(\xi^3+5\xi)\ (\Sigma\lambda_i^{-3} T_i^{-4})\,/\ (\Sigma\lambda_i^{-1} T_i^{-2})^{3/2} \\ &\quad + 18\xi\,(\Sigma\lambda_i^{-2} T_i^{-2})/(\Sigma\lambda_i^{-1} T_i^{-2})^{1/2}]\,(36n^{3/2})^{-1} \qquad (2.6)\end{aligned}$$

Upper confidence limits for $\theta$ are obtained by using negative values of $\xi$.

Table 3 shows some lower confidence limits for series systems and exponential times to failure ($t_m = 1$). Optimum here means uniformly most accurate unbiased – see Lentner and Buehler (1963). M.G. is the approximately optimum method of Mann and Grubbs (1972). Further results are given in Table 3 of Winterbottom (1980).

There is some indication in Table 3 that (2.6) becomes superior to the approximately optimum method of Mann and Grubbs as $n$ increases and, of course, this is the region where calculations of confidence limits become very difficult and for which the asymptotic expansion is guaranteed to become more accurate.

An alternative means of demonstrating the effectiveness of asymptotic expansions is to generate component test results with given sample sizes when the component reliabilities and hence the system reliability are specified. For each set of component test results suppose a lower limit is calculated using the appropriate asymptotic expansion. This procedure is repeated a large number of times with the sample sizes and parameters fixed. The proportion of lower limits falling below the true system reliability estimates the achieved confidence level.

Table 4 shows the results of such a Monte-Carlo study for two systems, a series-parallel system (S.P.) and a system with standby redundancy (S.R.). For

**Table 3 Lower confidence limits for series systems and exponential times to failure**

| $\gamma$ | $(T_i, n_i)$ | Optimum | (2.6) | M.G. |
|---|---|---|---|---|
| 0.9 | (17.9855,2)(7.305,2)(31.271,2) | 0.731 | 0.734 | 0.734 |
| 0.9 | (15.945,2)(15,2637,3)(10.688,4) | 0.772 | 0.771 | 0.771 |
| 0.95 | (4.375,4)(12.95,4)(5.475,4) | 0.509 | 0.509 | 0.511 |
| 0.95 | (10.1,4)(2.825,4)(7.574,4) | 0.417 | 0.419 | 0.424 |

**Table 4 Series-parallel and standby redundancy systems simulated achieved confidence levels: 5000 trials in each case**

| | $n$ | $\theta_1$ | $\theta_2$ | $\psi(\theta_1,\theta_2)$ | 0.9 level | 0.975 level | 0.99 level |
|---|---|---|---|---|---|---|---|
| S.P.(N+2) | 2 | 16 | 8 | 0.9264 | 0.903 | 0.977 | 0.988 |
| | 2 | 32 | 32 | 0.9683 | 0.901 | 0.975 | 0.990 |
| S.R.(N+1) | 10 | 5 | 0.96 | 0.9759 | 0.906 | 0.977 | 0.988 |

S.P. all components have exponential times to failure and the reliability function is $\psi(\theta_1,\theta_2) = \exp(-\theta_1^{-1})\{2\exp(-\theta_2^{-1})-\exp(-2\theta_2^{-1})\}$. The sample sizes used were $n_1 = 2$, $n_2 = 4$ and since $n = 2$ is small two adjustments were used, denoted by S.P. $(N+2)$. The reliability function for the standby system is $\psi(\theta_1,\theta_2) = \exp(-\theta_1^{-1})(1+\theta_2/\theta_1)$. Here $\theta_1$ is the exponential mean time to failure for the two components, one active and one standby. If the first component fails before mission time $t_m = 1$ the second component is switched in and $\theta_2$ is the probability that the switch works. $n_1 = 10$ and $n_2 = 50$ so that $n = 10$ and only one adjustment was used, denoted by S.R. (N+1). Note that the achieved confidence levels are close to the designated levels even for the extreme case $\gamma = 0.99$.

## 3. EXPANSIONS FOR BAYESIAN INTERVALS

Before discussing the approach for systems with general structures we show how percentiles of the posterior distributions of series system reliability can be well approximated using Cornish and Fisher expansions.

For a series system of $k$ independent components let $k_1$ be tested on a pass/fail basis and suppose that a test procedure based on exponential times to failure is used for the remaining $k_2 = k - k_1$ components. The system reliability for mission time $t_m$ is given by

$$R = \prod_{i=1}^{k_1} R_i \cdot \exp(-t_m \sum_{j=1}^{k_2} \Lambda_j) ,$$

where $R_i (i=1,2,\ldots,k_1)$ is the reliability of the $i$th component and $\Lambda_j$ $(j=1,2,\ldots,k_2)$ is the failure rate of the jth component. Using well known results for conjugate families suppose the posterior distributions for the $R_i$ are beta with

$$f_i(r_i) = \frac{r_i^{\alpha_i-1}(1-r_i)^{\beta_i-1}}{B(\alpha_i,\beta_i)}, \quad 0 < r_i < 1 ,$$

and for the $\Lambda_j$ the posterior distributions are gamma with

$$g_j(\lambda_j) = \tau_j(\tau_j\lambda_j)^{\eta_j-1} e^{-\tau_j\lambda_j}/\Gamma(\eta_j), \lambda_j > 0 .$$

Without loss of generality we may take the mission time $t_m = 1$ and then taking the negative of the natural logarithm of system reliability gives

$$Y = \sum_{i=1}^{k_1} Y_i + \sum_{j=1}^{k_2} \Lambda_j ,$$

where $Y = -lnR$ and $Y_i = -lnR_i$.

The Laplace transform of the probability density function of $Y$ is

$$E\{\exp(-sY)\} = \prod_{i=1}^{k_1} \left\{ \frac{\alpha_i(\alpha_i+1)\ldots(\alpha_i+\beta_i-1)}{(\alpha_i+s)(\alpha_i+s+1)\ldots(\alpha_i+s+\beta_i-1)} \right\} \prod_{j=1}^{k_2} \left( \frac{\tau_j}{\tau_j+s} \right)^{\eta_j}. \tag{3.1}$$

From the Laplace transform it is straightforward to show that the cumulants for the distribution of $Y$ are given by

$$\kappa_r = (r-1)!\{\sum_{i=1}^{k_1} \sum_{u=0}^{\beta_i-1} (\alpha_i+u)^{-r} + \sum_{j=1}^{k_2} \eta_j \tau_j^{-r}\}, \; r = 1, 2, \ldots \tag{3.2}$$

Using the notation in Kendall and Stuart (1963) we have $l_1 = 0$, $l_2 = 0$ and $l_r = \kappa_r / (\kappa_2)^{r/2}$, $r > 2$. Then approximate percentiles of the distribution of $Y$ are given by the Cornish and Fisher expansion

$$\begin{aligned} Y_\xi \simeq \kappa_1 + \kappa_2^{1/2}[\xi &+ \{l_3(\xi^2-1)/6\} + \{l_4(\xi^3-3\xi)/24 - l_3^2(2\xi^3-5\xi)/36\} \\ &+ \{l_5(\xi^4-6\xi^2+3)/120 - l_3l_4(\xi^4-5\xi^2+2)/24 \\ &+ l_3^3(12\xi^4-53\xi^2+17)/324\} \\ &+ \{l_6(\xi^5-10\xi^3+15\xi)/720 - l_4^2(3\xi^5-24\xi^3+29\xi)/384 \\ &- l_3l_5(2\xi^5-17\xi^3+21\xi)/180 + l_3^2l_4(14\xi^5-103\xi^3+107\xi)/288 \\ &- l_3^4(252\xi^5-1688\xi^3+1511\xi)/7776\}] , \end{aligned} \tag{3.3}$$

where $\xi$ is an appropriate standard normal value corresponding to the desired percentile. By the Central Limit Theorem we know that for large values of $k = k_1 + k_2$ the distribution of $Y$ is approximately normal and this leads to the large sample formula $Y_\xi \simeq \kappa_1 + \kappa_2^{1/2}\xi$. When multiplied by $\kappa_2^{1/2}$ the four expressions in braces in (3.3) provide adjustments of successively higher order to the large sample formula. The use of four adjustments yields an accuracy of approximation typified by the following example.

*Example* 1. Let $k = 3$ with $k_1 = 2$ and $k_1 = 1$. Suppose $\alpha_1 = 18$, $\beta_1 = 2$; $\alpha_2 = 19$, $\beta_2 = 1$; $\tau_1 = 40$, $\eta_1 = 3$. Inverting the Laplace Transform (3.1) to obtain the probability density function of $Y$ and then integrating gives the exact distribution function

$$\begin{aligned} H(r) = 2169.79714r^{18} &- 2150.21518r^{19} - 2363.4567(-lnr)r^{19} \\ &- 18.58196r^{40} - 177.55857(-lnr)r^{40} \\ &- 535.80705(-lnr)^2r^{40} . \end{aligned}$$

Then, for example, the end points of a two-sided 95% Bayesian interval for series system reliability are the solutions of $H(r) = 0.025$ and $H(r) = 0.975$. These values are 0.6201 and 0.9208 respectively. In order to use (3.3) we first calculate the cumulants and then the ratios $l_r, r > 2$. From (3.2) we obtain

$$\kappa_r = (r-1)!\{18^{-r} + 2\times 19^{-r} + 3\times 40^{-r}\}\ ,\ r = 1,2,\ldots$$

and thus $\kappa_1 = 0.23581871$, $\kappa_2 = 0.010501586$, $l_3 = 0.94767322$, $l_4 = 1.41696416$, $l_5 = 2.90135699$ and $l_6 = 7.52703846$. For $\xi = \pm 1.95996$ we first calculate from (3.3) the values of $Y_\xi$ and then $R_\xi = \exp(-Y_\xi)$. These values are 0.6201 and 0.9209, the lower limit for $R$ being obtained from the upper limit for $Y$ and vice versa. Table 5 shows the same high accuracy for a variety of cases where exact (E.) lower limits have been compared with corresponding lower limits given by the asymptotic expansion(A.E.)

For general structures a possible approach is to adapt (2.1). We now write

$$Z = n^{1/2}\{g_1/n + (h + g_2/n)(R-\rho) + g_3(R-\rho)^2 + g_4(R-\rho)^3\}$$

where $\rho = \psi(\rho_1, \rho_2, \ldots, \rho_k)$ and $R = \psi(R_1, R_2, \ldots, R_k)$. Here the pseudo parameter $\rho_i$ could be the mean of the posterior distribution of $R_i, i = 1, 2, \ldots, k$. Then take $h, g_1, \ldots, g_4$ as functions of $\rho_1, \ldots, \rho_k$ and of $\lambda_1, \ldots, \lambda_k$ only. First it will be necessary to establish the cumulant property in the $n_i$ so that the methodology of Section 2 can be applied. The algebra should be somewhat less extensive than for the classical approach and it may be possible to derive general expressions enabling a third adjustment to be made. Not the distinction between the series case given in this section where the asymptotic variable is $k$ and the general approach where the asymptotic variable is $n$. Both approaches can be applied to series systems and the comparative results should be interesting. In order to assess the effectiveness of the resulting expansions it will be necessary to simulate the posterior distributions of $R$ for a variety of structures and test data.

**Table 5 Comparisons for lower limits**

| k | | | | Level 0.9 | 0.95 | 0.975 |
|---|---|---|---|---|---|---|
| 2 | $\alpha_1 = 2$ | $\beta_1 = 1$ | E. | .1770 | .1239 | .0871 |
| | $\alpha_2 = 4$ | $\beta_2 = 1$ | A.E. | .1768 | .1237 | .0871 |
| 2 | $\alpha_1 = 30$ | $\beta_1 = 1$ | E. | .8774 | .8555 | .8349 |
| | $\alpha_2 = 50$ | $\beta_2 = 2$ | A.E. | .8774 | .8557 | .8351 |
| 3 | $\alpha_1 = 11$ | $\beta_1 = 1$ | E. | .6992 | .6544 | .6134 |
| | $\tau_1 = 20$ | $\eta_1 = 1$ | A.E. | .6991 | .6544 | .6137 |
| | $\tau_2 = 30$ | $\eta_2 = 2$ | | | | |
| 4 | $\alpha_1 = 11$ | $\beta_1 = 1$ | E. | .6195 | .5731 | .5320 |
| | $\alpha_2 = 12$ | $\beta_2 = 1$ | A.E. | .6195 | .5731 | .5319 |
| | $\tau_1 = 20$ | $\eta_1 = 1$ | | | | |
| | $\tau_2 = 30$ | $\eta_2 = 2$ | | | | |

## REFERENCES

Cornish, E.A. and Fisher, R.A. (1937). Moments and cumulants in the specification of distributions. *Int. Statist. Rev.* **5**, 307-22.

James, G.S. (1955). Cumulants of a transformed variate. *Biometrika* **42**, 529-31.

James, G.S. and Mayne, A.J. (1962). Cumulants of functions of random variables. *Sankhyā* A **24**, 47-54.

Kendall, M.G. and Stuart, A. (1963). *The Advanced Theory of Statistics*, Vol. 1, 166 (Griffin and Co. Ltd., London).

Lentner, M.M. and Buehler, R.J. (1963). Some inferences about gamma parameters with an application to a reliability problem. *J. Am. Statist. Assoc.* **58**, 670-7.

Lipow, M. and Riley, J. (1960). *Tables of upper confidence limits on failure probability of 1, 2 and 3 component serial systems*, Vols. 1 and 2. Space Technology Laboratories National Technical Information Service, AD-609-100, AD-636-718, Clearinghouse, U.S. Dept. of Commerce.

Mann, N.R. and Grubbs, F.E. (1972). Approximately optimum confidence bounds on series system reliability for exponential time to fail data. *Biometrika* **59**, 191-204.

Winterbottom, A. (1980). Asymptotic expansions to improve large sample confidence intervals for system reliability. *Biometrika* **67**, 351-7.

# Confidence Bounds Based on Sample Orderings

M. Vernon Johns, Jr.*

*Stanford University*

**Abstract**

The basic problem of determining objective (frequentistic) confidence bounds for the reliability of a series system based on failure data from tests of the independent components is addressed. The notion of confidence bounds based on orderings imposed on the sample space is exploited, and certain optimality considerations are incorporated. Examples of exact confidence bounds are produced for the case of three-component systems. These bounds are computed using sample orderings generated sequentially by a two-stage, "look ahead" optimization procedure.

## 1. INTRODUCTION

The problem of constructing confidence bounds for specified functions of several parameters is often rather intractable. The assessment of reliability for multicomponent systems based on component test data frequently involves the need for such confidence bounds. Situations of this type rarely admit the existence of uniformly most accurate confidence bounds (Lehmann (1959)), since the required families of uniformly most powerful tests usually do not exist. Thus for such problems it may be reasonable to seek confidence bound procedures not motivated by hypothesis testing considerations.

Such an alternative approach has been suggested by Beuhler (1957) in which a connection between confidence bounds and specified total orderings of the sample outcomes is introduced and exploited in the context of a two-component parallel system with binomial test data. This approach has subsequently been applied to specific system reliability problems by several authors,

---

*This work was supported by U.S. Army Research Office Contract No. DAAG29-79-C-0166. The presentation of the material in Section 2 has benefitted from a critical reading by Robert M. Bell.

e.g., Harris and Soms (1980), Johns (1977, 1981), Johnson (1969), Lipow and Riley (1959), and Winterbottom (1974). The purpose of the present study is to discuss the construction of exact confidence bounds based on sample orderings and to explore some of the properties of such bounds in a context general enough to include most potential applications.

In the following the usual terminology for "exact" confidence bounds is employed. That is, a real valued function $b(x)$ is said to be an exact $1 - \alpha$ upper confidence bound for the function $\psi(\theta)$ of the parameter $\theta$ if

$$P_\theta\{\psi(\theta) \leqslant b(X)\} \geqslant 1 - \alpha, \tag{1.1}$$

for all $\theta$, where $X$ represents the observations(s). Equivalently, $b(x)$ may be described as an exact confidence bound "at level $1 - \alpha$" or "with confidence coefficient $1 - \alpha$." In Section 2 the properties of such confidence bounds based on specified sample orderings are developed in a series of propositions, and specializations to specific reliability problems are indicated. In Section 3 a numerical example is discussed.

## 2. BOUNDS AND ORDERINGS

As was indicated in Section 1, the idea of using sample orderings to generate confidence bounds was first introduced by Buehler (1957) who discussed the validity of the proposed method in the context of a specific reliability problem. Bol'shev and Loginov (1969) discuss the construction of confidence bounds monotone in the sample orderings generated by certain functions of the observations. In the following, which is a revision and extension of Johns (1975), we develop the theory with emphasis on the sample orderings themselves rather than possible generators of the orderings. The intent here is to present the central ideas in a framework general enough to include most statistical applications while avoiding as much as possible considerations special to particular problems.

To develop the general ideas relating exact confidence bounds to sample orderings it is convenient to introduce a fairly abstract statistical model. Let the sample space $\Xi$ be endowed with a measurable total ordering relation "$\overset{\circ}{\leqslant}$" and let $X$ represent the random outcome of the experiment where the space of possible outcomes is $\Xi$. Suppose that the possible distributions of $X$ are determined by the family of probability measures $P_\theta$, indexed by $\theta$, an element of the parameter space $\Theta$. Our objective is to find a $1 - \alpha$ level upper confidence bound for a specified real-valued function $\psi(\theta)$ defined on $\Theta$ where the range $R$ of $\psi(\theta)$ is assumed to be closed and bounded below. The quantity $\alpha \in (0, 1)$ is regarded as fixed throughout. We make the following definitions and assumptions:

*Definition D1.* For each $x \in \Xi$ let

$$S_\alpha(x) = \{\theta: P_\theta\{X \overset{\circ}{\leqslant} x\} > \alpha\}. \tag{2.1}$$

*Definition D2.* For each $x \in \Xi$ let

$$b(x) = \begin{cases} \sup\{\psi(\theta): \theta \in S_\alpha(x)\}, \text{ if } S_\alpha(x) \text{ is nonempty,} \\ \inf R, \text{ otherwise.} \end{cases} \tag{2.2}$$

*Remark 1.* By D2 if $\psi(\theta) > b(x)$, then necessarily $P_\theta\{X \overset{\circ}{\leqslant} x\} \leqslant \alpha$.

*Assumption A1.* For every subset $C$ of $\Xi$ having the property that if $x \in C$ and $y \overset{\circ}{\leqslant} x$ then $y \in C$, there exists an ordered sequence $x_1 \overset{\circ}{\leqslant} x_2 \overset{\circ}{\leqslant} \cdots$ of elements of $C$ such that $C = \bigcup_{n=1}^{\infty}\{x: x \overset{\circ}{\leqslant} x_n\}$.

*Assumption A2.* For each $x \in \Xi$, if $\psi(\theta) = b(x)$, then $P_\theta\{X \overset{\circ}{\leqslant} x\} \leqslant \alpha$.

*Remark 2.* By D1, D2, and A2 if $\psi(\theta) = b(x)$, then $\theta \notin S_\alpha(x)$, i.e., the supremum in D2 is never attained.

We now establish the following propositions.

*Proposition P1.* The function $b(x)$ is monotone in the ordering on $\Xi$.

*Proof:* If $x, y \in \Xi$ and $x \overset{\circ}{\leqslant} y$, then $S_\alpha(x) \subset S_\alpha(y)$ (D1) which in turn implies $b(x) \overset{\circ}{\leqslant} b(y)$ (D2). □

*Proposition P2.* Under assumptions A1 and A2 the function $b(x)$ is an upper confidence bound for $\psi(\theta)$ at level $1 - \alpha$. In particular,

$$P_\theta\{\psi(\theta) < b(X)\} \geqslant 1 - \alpha \quad \text{for all } \theta \in \Theta. \tag{2.3}$$

*Proof:* For arbitrary $\theta_0 \in \Theta$, let $\psi_0 = \psi(\theta_0)$ and $C_0 = \{x: b(x) \leqslant \psi_0\}$. The result follows immediately for all $\theta_0$ for which $C_0$ is empty. Assume that $C_0$ is nonempty. Then by P1 the set $C_0$ possesses the property required in A1 for the existence of a sequence $\{x_n\} \subset C_0$ such that $x_n \overset{\circ}{\leqslant} x_{n+1}$ for all $n$, and $C_0 = \bigcup_{n=1}^{\infty} \{x{:}x \overset{\circ}{\leqslant} x_n\}$. This implies that, as $n \to \infty$,

$$P_{\theta_0}\{X \overset{\circ}{\leqslant} x_n\} \uparrow P_{\theta_0}\{X \in C_0\}. \tag{2.4}$$

But by Remark 1 and A2, for all $n$, $P_{\theta_0}\{X \overset{\circ}{\leqslant} x_n\} < \alpha$. Hence $P_{\theta_0}\{X \in C_0\} \leqslant \alpha$ and the desired result follows. □

*Proposition P3.* Under assumptions A1 and A2, if $\tilde{b}(x)$ is any $R$-valued confidence bound such that $P_\theta\{\psi(\theta) < \tilde{b}(X)\} \geqslant 1 - \alpha$ for all $\theta \in \Theta$, then

i(i) $\sup_{y \overset{\circ}{\leqslant} x} \tilde{b}(y) \geqslant b(x)$ for all $x \in \Xi$, and

(ii) if $\tilde{b}(x)$ is monotone in the ordering on $\Xi$, then $\tilde{b}(x) \geqslant b(x)$ for all $x \in \Xi$.

*Proof:* First we assume that $\tilde{\mathbf{b}}(x)$ is monotone and establish (ii). Suppose there exists an $x' \in \Xi$ such that $\tilde{b}(x') < b(x')$. Then $S_\alpha(x')$ must be nonempty and by Remark 2 following A2, the sup defining $b(x')$ is not attained. Hence there exists a $\theta' \in S_\alpha(x')$ such that $\tilde{b}(x') < \psi(\theta') < b(x')$, and $P_{\theta'}\{X \overset{\circ}{\leqslant} x'\} > \alpha$. Thus by the monotonicity of $\tilde{b}(x)$,

$$P_{\theta'}\{\tilde{b}(X) \leqslant \psi(\theta')\} \geqslant P_{\theta'}\{\tilde{b}(X) \leqslant \tilde{b}(x')\} = P_{\theta'}\{X \overset{\circ}{\leqslant} x'\} > \alpha. \quad (2.5)$$

This contradicts the hypothesis that $P_\theta\{\psi(\theta) < \tilde{b}(X)\} \geqslant 1 - \alpha$ for all $\theta$ and establishes (ii). To show (i) we let $b^*(x) = \sup_{y \leqslant x} \tilde{b}(y)$ and observe that $b^*(x)$ is monotone in the ordering on $\Xi$ and $b^*(x) \geqslant \tilde{b}(x)$ for all $x \in \Xi$. Hence if $\tilde{b}(x)$ is a $1 - \alpha$ confidence bound for $\psi(\theta)$, so is $b^*(x)$ and applying (ii) to $b^*(x)$ yields (i). □

In order to specialize these results in the direction of applications we henceforth assume that the parameter $\theta$ is of dimension $k$, i.e., $\theta = (\theta_1, \theta_2, \ldots, \theta_k)$ where the $\theta_i$'s are real. Without essential loss of generality we assume that $\Theta$ contains the positive orthant. Within this framework we make the following additional assumptions:

*Assumption A3.* The function $\psi(\theta)$ is continuous and strictly increasing in each of the $\theta_i$'s.

*Assumption A4.* For any $x \in \Xi$, $P_\theta\{X \overset{\circ}{\leqslant} x\}$ is continuous in each of the $\theta_i$'s.

*Proposition P4.* Assumptions A3 and A4 imply that Assumption A2 is satisfied.

*Proof:* Suppose that for some $x' \in \Xi$ there exists a $\theta' \in \Theta$ such that $\psi(\theta') = b(x')$ and $P_{\theta'}\{X \overset{\circ}{\leqslant} x'\} > \alpha$. They by A3 and A4 we can find a $\theta'' \in \Theta$ with $\theta_i'' \geqslant \theta_i'$ for all $i$ and $\theta_{i_0}'' > \theta_{i_0}''$ for some $i_0$ such that $\psi(\theta'') > b(x')$ and $P_{\theta''}\{X \overset{\circ}{\leqslant} x'\} > \alpha$. This contradicts D2 and the result follows. □

We now establish two corollaries to Proposition P4 which permit applications of the foregoing results to the construction of confidence bounds for the reliability of series systems.

*Corollary C1.* If $X = (X_1, X_2, \ldots, X_k)$ where the $X_i$'s which represent observed numbers of failures are independent binomially distributed random variables with sample sizes $n_1, n_2, \ldots, n_k$ and probabilities of component failure $\theta_1, \theta_2, \ldots, \theta_k$ respectively, then letting $\psi(\theta) = 1 - \prod_{i=1}^k (1 - \theta_i)$ we conclude that $b(x)$ given by D2 is an upper confidence bound for $\psi(\theta)$ at level $1 - \alpha$ and hence $1 - b(x)$ is a lower confidence bound for the system reliability $\prod_{i=1}^k (1 - \theta_i)$.

*Corollary C2.* If (i) $X = (X_1, X_2, \ldots, X_k)$ where the $X_i$'s are independent Poisson random variables with parameters $\theta_1, \theta_2, \ldots, \theta_k$ respectively, and (ii) $\psi(\theta) = a_1\theta_1 + a_2\theta_2 + \cdots + a_k\theta_k$ where the $a_i$'s are positive, then $b(x)$ given by D2 is an upper confidence bound for $\psi(\theta)$ at level $1 - \alpha$.

*Proof of C1 and C2:* A1 is satisfied for any ordering since $X$ is discrete. A3 is clearly satisfied for the forms of $\psi(\theta)$ given in C1 and C2, and A4 is satisfied for binomial and Poisson random variables. The desired results follow from P2 and P4. □

Corollary C1 generalizes at once to coherent systems by taking $\psi(\theta) = 1 - r(\theta_1, \theta_2, \ldots, \theta_k)$ where $r(\cdot)$ is the reliability function of the $k$ component system.

The structure considered in Corollary C2 arises in connection with the use of the Poisson approximation to the binomial distribution which is appropriate when dealing with highly reliable series systems. In this situation the system reliability is approximated by $1 - c\psi(\theta)$ where $c = \sum_{i=1}^{k} 1/n_i$ and $a_i = 1/cn_i$, $i = 1, 2, \ldots, k$, as discussed in Johns (1981). The lower confidence bound for system reliability then becomes $1 - cb(x)$, and the form of $\psi(\theta)$ makes the construction of useful tables feasible.

The actual computation of the bound $b(x)$ given by D2 is greatly facilitated if $P_\theta\{X \overset{\circ}{\leqslant} x\}$ is monotone in the components of $\theta$. The following proposition gives conditions guaranteeing this property:

*Assumption A5.* The components of $X = (X_1, X_2, \ldots, X_k)$ are independent, and the distribution of each $X_i$ depends only on the corresponding $\theta_i$ and is stochastically nondecreasing in $\theta_i$.

*Definition D3.* We denote by $(\overset{*}{\leqslant})$ the natural partial ordering of $X$ generated by componentwise dominance. That is, $x \overset{*}{\leqslant} y$ if and only if $x_i \leqslant y_i$ for $i = 1, 2, \ldots, k$, with $x \overset{*}{<} y$ if at least one of these inequalities is strict. An arbitrary total ordering $(\overset{\circ}{\leqslant})$ on $\Xi$ is *consistent* with $(\overset{*}{\leqslant})$ if $x \overset{*}{\leqslant} y$ implies $x \overset{\circ}{\leqslant} y$.

*Proposition P5.* If the ordering on $\Xi$ is consistent with the natural partial ordering and Assumption A5 is satisfied, then for each $x \in \Xi$, $P_\theta\{X \overset{\circ}{\leqslant} x\}$ is nonincreasing in each component of $\theta$.

*Proof:* For $y \in \Xi$ we introduce the representation $y = (y_1, y^{(2)})$ where $y^{(2)} = (y_2, y_3, \ldots, y_k)$. For real $z$, fixed $x \in \Xi$, and all $y \in \Xi$, let $I_x(z, y^{(2)})$ be the indicator function of the set $\{y^{(2)}: (z, y^{(2)}) \overset{\circ}{\leqslant} x\}$. Then

$$P_\theta\{X \overset{\circ}{\leqslant} x\} = E_\theta\, I_x(X_1, X^{(2)}). \tag{2.6}$$

For any $y \in \Xi$, if $z' \leqslant z''$, then $(z', y^{(2)}) \overset{\circ}{\leqslant} (z'', y^{(2)})$, by the consistency hypothesis, and $I_x(z', y^{(2)}) \geqslant I_x(z'', y^{(2)})$. Hence letting $G_x(z) = E_\theta\{I_x(X_1, X^{(2)}) \mid X_1 = z\}$ we see that $G_x(z)$ is nonincreasing in $z$. Thus, since the distribution of $X_1$ is stochastically nondecreasing in $\theta_1$ (A5), we conclude that $E_\theta\, I_x(X_1, X^{(2)}) = E_\theta\, G_x(X_1)$ is nonincreasing in $\theta_1$. The same argument applies to the other components of $\theta$ establishing the desired result. □

Suppose that $\Theta$ is the nonnegative orthant of $R^{(k)}$ and let $\Theta_0$ be the simplex $\Theta_0 = \{\theta: \sum_{i=1}^{k} \theta_i = 1\}$. If A4 and A5 are satisfied, then by P5 we observe that for any $\theta \in \Theta_0$ and real $d$, $P_{d\theta}\{X \overset{\circ}{\leqslant} x\}$ is continuous and nonincreasing in $d$. If the lower bound of $P_{d\theta}\{X \overset{\circ}{\leqslant} x\}$ as $d \to \infty$ is less than $\alpha$ for all $x \in \Xi$ and all $\theta \in \Theta_0$, then for each $x \in \Xi$ and $\theta \in \Theta_0$ there exists a smallest number $d = d(x, \theta)$ such that $P_{d(x,\theta)\theta}\{X \lesseqgtr x\} = \alpha$. The confidence bound $b(x)$ defined by D2 is then given by

$$b(x) = \sup_{\theta \in \Theta_0} \psi(d(x, \theta)\theta). \tag{2.7}$$

Now $d(x, \theta)$ is easily computed using root-finding techniques so that the computation of $b(x)$ reduces to searching over $\Theta_0$ for the maximum of $\psi(d(x, \theta)\theta)$. Many routines are available for implementing such searches. For the situations described in Corollaries C1 and C2, the values of $d$ such that $P_{d\theta}\{X \lesseqgtr x\} = \alpha$ are unique and (2.7) is a computationally feasible version of (1.5).

All of the above results apply *mutatis mutandis* to the construction of *lower* confidence bounds and hence confidence intervals.

For situations such as those arising in many reliability applications for which no uniformly most accurate confidence bounds exist, the problem of finding suitable sample orderings becomes central. In this connection several methods have been proposed and explored to some extent in Johns (1981). These procedures for ordering the sample points are of the following three general types:

(i) The sample space is ordered by the ordered values of a function $\hat{\psi}(x)$ where $\hat{\psi}$ is either the maximum likelihood estimate of $\psi(\theta)$ or the asymptotic maximum likelihood confidence bound for $\psi(\theta)$. The use of such functions which are usually comparatively simple in form guarantees that the corresponding confidence bound $b(x)$ will at least be asymptotically optimal. For the case covered by corollary C2, for example, the two versions of $\hat{\psi}$ are $\hat{\psi} = \sum_{i=1}^{k} a_i X_i$ and $\hat{\psi} = \sum_{i=1}^{k} a_i X_i + z_\alpha \left\{ \sum_{i=1}^{k} a_i^2 X_i \right\}^{1/2}$ respectively, where $z_\alpha$ is the upper $\alpha$-th quantitle of the standard normal distribution.

(ii) The sample space may be ordered so as to minimize the expected size of the confidence region with respect to some prior distribution over the values of $\theta$. For the case of an upper confidence bound $b(x)$ for $\psi(\theta)$ one could thus choose the ordering to minimize $E_H\, b(X)$ where $H$ is some prior distribution on $\Theta$. Such an approach at least guarantees the admissibility of the resulting bound. This method does not, of course, impair the frequentistic interpretation of the resulting confidence bounds.

(iii) When the sample space is discrete and possesses a natural "smallest" element, e.g., zero failures observed for each component in the reliability problem, it is possible to order the sample points sequentially according to various simple criteria. Thus, for the upper confidence bound problem, at stage $n$, one may compute the values of $b(x)$ for all sample points which are candidates for the $(n + 1)^{st}$ point in the ordering (consistent with the natural partial ordering) and select the point giving the smallest value of $b(x)$. This method can be generalized by looking ahead more than one step at each stage.

Among these three methods it appears (see Johns (1981)) that method (i) is the easiest to implement and method (ii) the hardest. The bounds produced by method (i) are, however, often somewhat larger than those produced by methods (ii) and (iii). Method (ii) can also be implemented sequentially, but the computational difficulties are formidable. Method (iii), with a two-step look ahead at each stage, was used in Johns (1981) to generate tables of $b(x)$ for $k = 2$ for the Poisson problem of Corollary C2. These results were found to agree well with a limited implementation of method (ii) using suitable prior distributions. An example of the application of method (iii) is given in the next section.

## 3. AN EXAMPLE

Consider a series system consisting of three independent components. Suppose that the components are tested independently with the respective numbers of trails, $n_1 = 70$, $n_2 = 50$, and $n_3 = 30$, and with the corresponding observed numbers of component failures represented by $x_1$, $x_2$, and $x_3$. The binomial model of Corollary C1 of Section 2 applies; and since the condition of Proposition P5 are also satisfied, the computational method expressed in (2.7) may be employed to obtain $b(x)$ and hence the lower confidence bound for the system reliability. Starting with the initial sample point $(x_1, x_2, x_3) = (0, 0, 0)$, the first ten sample points in the optimal sequentially determined ordering obtained by the two-step look-ahead method (iii) described in Section 2 are given in Table 1, together with the corresponding values of the exact 95% ($\alpha = .05$) lower confidence bounds for system reliability. For purposes of comparison the approximate 95% bounds obtained using the simple algorithm and the tables of Johns (1981) based on the Poisson approximation are also shown as well as the Lindstrom-Madden approximate 95% bounds.

Examination of Table 1 suggests that the Poisson approximation method is quite satisfactory for this example, at least for sample points appearing early in the ordering and thus associated with high indicated reliability. The Lindstrom-Madden bounds appear to be slightly too conservative.

**Table 1**

| Stage | Number of Observed Failures | | | Exact 95% Confidence Bound (binomial) | Approximate 95% Confidence Bound (Poisson) | Lindstrom-Madden 95% Confidence Bound |
|---|---|---|---|---|---|---|
| | $x_1$ | $x_2$ | $x_3$ | | | |
| 1 | 0 | 0 | 0 | .905 | .900 | .905 |
| 2 | 1 | 0 | 0 | .901 | .896 | .882 |
| 3 | 0 | 1 | 0 | .895 | .889 | .873 |
| 4 | 1 | 1 | 0 | .891 | .895 | .850 |
| 5 | 2 | 0 | 0 | .888 | .884 | .859 |
| 6 | 2 | 1 | 0 | .881 | .881 | .831 |
| 7 | 3 | 0 | 0 | .876 | .871 | .838 |
| 8 | 0 | 2 | 0 | .875 | .867 | .842 |
| 9 | 1 | 2 | 0 | .868 | .862 | .823 |
| 10 | 3 | 1 | 0 | .864 | .863 | .811 |

The exact binomial confidence bounds were found by computing $b(x)$ for all candidate points at each stage using an interactive computer program which calculates the quantity $d(x, \theta)$ appearing in (2.7) for a variable grid of $\theta$-points in the simplex $\Theta_0$ under operator control. For this particular example it developed that the first ten ordered sample points obtained by the rather tedious two-step look-ahead sequential procedure were the same as would have been obtained using only a one-step procedure. Experience with other examples indicates, however, that if the procedure were continued beyond ten stages, instances would arise where the results for the two-step method would represent slight improvements on the one-step results.

The Lindstrom-Madden bounds were calculated by the method described in Lloyd and Lipow (1977) using linear interpolation on numbers of failures in computing the one-dimensional binomial confidence bounds required by this method.

## REFERENCES

Bol'shev, L.N. and E.A. Loginov (1969), "Interval Estimates in the Presence of Noise," *Theory of Probability and Its Applications*, 11, 82-94.

Buehler, R.J. (1957), "Confidence Intervals for the Product of Two Binomial Parameters," *Journal of the American Statistical Association*, 52, 482-493.

Harris, B. and A.P. Soms (1980), "Bounds for Optimal Confidence Limits for Series Systems," Technical Summary Report No. 2093, University of Wisconsin-Madison, Mathematics Research Center.

Johns, M.V., Jr. (1975), "Reliability Assessment for Highly Reliable Systems," Technical Report No. 1, Stanford University, Department of Statistics.

Johns, M.V., Jr. (1977), "Some Aspects of Reliability Assessment Based on Sample Orderings," *Proceedings of the ARO Workshop on Reliability and Probabilistic Design*, 101-110.

Johns, M.V., Jr. (1981), "Feasible Objective Confidence Bounds for System Reliability," Technical Report No. 6, Stanford University, Department of Statistics.

Johnson, J.R. (1969), "Confidence Interval Estimation of the Reliability of Multicomponent Systems Using Component Test Data," Ph.D. dissertation, University of Delaware.

Lehmann, E.L. (1959), *Testing Statistical Hypotheses*, New York: John Wiley and Sons.

Lipow, M. and J. Riley (1959), "Tables of Upper Confidence Limits on Failure Probability of 1, 2 and 3 Component Serial Systems," Report No. TR-59-0000-00756, Space Technology Laboratories, Inc.

Lloyd, D.K. and M. Lipow (1977), *Reliability: Management, Methods, and Mathematics*, 2nd ed., Redono Beach: Published by the Authors.

Winterbottom, A. (1974), "Lower Limits for Series System Reliability from Binomial Data," *Journal of the American Statistical Association*, 69, 782-788.

# Multistate Systems

Henry W. Block and Thomas H. Savits

*University of Pittsburgh*

**Abstract**

The relationship between systems and subsystems or components is discussed in this paper. The usual context of this topic has been in the relationship between components which can be in only one of two modes, "on" or "off," and a system which is either "on" or "off." Traditional functional relationships describing this situation fail to take into account the problem of component redundance. Here we study the situation where components and the system can be described by a multiplicity of modes from "not functioning" to "functioning at peak capacity." Topics discussed are component relevance to the system and importance of components to the system. A technical aspect discussed is the decomposition of the functional form of the system, so that methods applicable to "on-off" systems can be applied.

## 1. "On"-"Off" SYSTEMS AND COMPONENTS

In attempting to describe the relationships between a system and its components, monotone structure functions have been used. The types of systems which have been described in this way satisfy the following assumptions:

(1) the system's operation is a function of the operation of its components each of which has only the two possible modes — "on" or "off;"
(2) if a component's performance is improved (i.e., if it goes from "off" to "on"), the system's performance cannot get worse (i.e., the system if "on" will not go "off");
(3) if all components are "off" ("on") the system is "off" ("on").

The mathematical form of these assumptions then becomes:

(1) the system's operation (0 for "off" and 1 for "on") is given by a function $\phi(x_1, x_2, \ldots, x_n)$ where $x_1$ describes the operation of the $i^{\text{th}}$ component (0 for "off" and 1 for "on");
(2) $\underline{x} \leqslant \underline{y}$ implies that $\phi(\underline{x}) \leqslant \phi(\underline{y})$, e.g. $\underline{x} = (0, 1, \ldots, 1) \leqslant (1, 1, \ldots, 1) = \underline{y}$ implies that if $\phi(\underline{x}) = 1$, then $\phi(\underline{y}) = 1$;
(3) $\phi(0, 0, \ldots, 0) = 0$ and $\phi(1, 1, \ldots, 1) = 1$.

An example is a 2 out of 3 system where the system is operating if at least 2 components are operating. This can be written as

$$\phi(x_1, x_2, x_3) = \begin{cases} 1 \text{ if } x_1 + x_2 + x_3 \geqslant 2, \\ 0 \text{ if } x_1 + x_2 + x_3 < 2. \end{cases}$$

Note that we can also represent $\phi$ in two other alternate ways:

$$\phi(x_1, x_2, x_3) = \max_{1 \leqslant j \leqslant 3} \min_{i \neq j} x_i = \min_{1 \leqslant j \leqslant 3} \max_{i \neq j} x_i.$$

The above two representations are called the min path and min cut representations, respectively, and are valid for every monotone structure function $\phi$: i.e., there exist min path sets $P_1, \ldots, P_p$ and min cut sets $K_1 \ldots, K_k$ such that

$$\phi(x_1, \ldots, x_n) = \max_{1 \leqslant j \leqslant p} \min_{i \in P_j} x_i = \min_{1 \leqslant j \leqslant k} \max_{i \in K_j} x_i.$$

Each min path set is a minimum set of components whose functioning insures the functioning of the system while each min cut set is a minimal set of elements whose failure causes the system to fail. These representations have proven to be very useful (cf. Barlow and Proschan (1975)).

The mathematical formulation of systems in this way has led to several developments. One of these is the making explicit of certain engineering concepts and another is the calculation of bounds for expected system lifetimes. An example of the former is the intuitive principle:

Redundancy at the component level is more effective than at the systems level. (1)

It can be shown mathematically that this principle holds for any monotone system of the type just described.

To illustrate the latter development (i.e., obtaining bounds) we consider the following which is defined for any $t \geqslant 0$.

$$X(t) = \begin{cases} 1 \text{ if the system is ``on'' at time } t, \\ 0 \text{ if the system is ``off'' at time } t, \end{cases}$$

and, for components $i = 1, 2, \ldots, n$,

$$X_i(t) = \begin{cases} 1 \text{ if component } i \text{ is "on" at time } t, \\ 0 \text{ if component } i \text{ is "off" at time } t. \end{cases}$$

We then have

$$X(t) = \phi(X_1(t), X_2(t), \ldots, X_n(t)).$$

Now let $T$ be the lifetime of the system and $T_i$ be the lifetime of the $i$th component for $i = 1, 2, \ldots, n$. We then have plotting $t$ versus $X(t)$ and similarly for $X_i(t)$. From this it is clear that $X(t_0) = 1$ if and only if $T > t_0$ for any fixed time $t_0 \geqslant 0$.

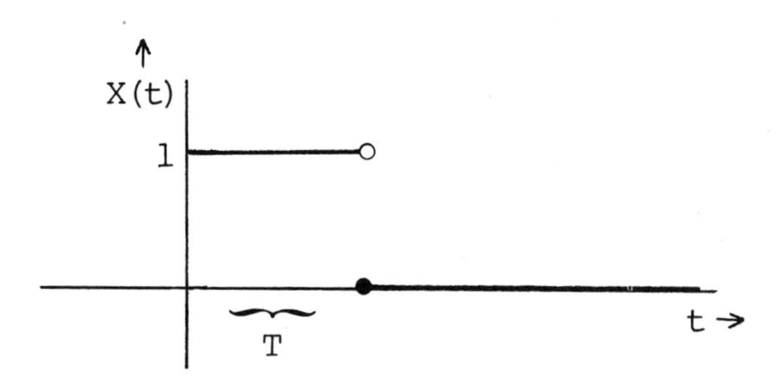


*Example 1.* It is shown that the lifetimes of a certain component (which is made with parts whose lifetimes have increasing hazard rates) has a probability of 0.95 of surviving 1000 hours. Then by the application of several results from mathematical reliability theory, if $T$ is the component lifetime, then

$$P\{T > t\} \geqslant \exp(-\alpha t) \text{ for } t \leqslant 1000$$

where $\alpha = -\log(.95)/1000$. In particular for $t = 900$ we have

$$P\{T > 900) \leqslant .955.$$

We also know $P\{T > t\} \leqslant \exp(-\alpha t)$ for $t \geqslant 1000$. (The results used in deriving this are the IFRA closure result and the single crossing property for these lifetimes). Thus from the information of one probability at a single times we can find conservative estimates on all probabilities at any time less than the single time and upper estimates for times greater than the single time.

Another use of these monotone structure functions is to mathematically describe which components in a system are most important to the system's operation. One way to measure this importance is by measuring the number of times a component contributes to a system's operation. We give a few examples to illustrate the idea of component importance.

*Example 2.* Consider the 2 out of 3 system described previously and consider the first component. Then $0 = \phi(0, 0, 1) < \phi(1, 0, 1) = 1$ so $(\cdot, 0, 1)$ contributes to the operation of component 1. A similar remark holds for $(\cdot, 1, 0)$. There are 4 vectors of this type of which only two contribute. So we can say component one has importance $\frac{2}{4} = \frac{1}{2}$. Similarly components 2 and 3 all have importance $\frac{1}{2}$. This gives that all three components are equally important as would be expected.

For a general system this measure of importance becomes

$$I_\phi(i) = \frac{1}{2^{n-1}} \#\{(\cdot_i, \underline{x}) \mid 0 = \phi(0_i, \underline{x}) < \phi(1_i, \underline{x}) = 1\},$$

where $(\cdot_i, \underline{x}) = (x_1, \ldots, x_{i-1}, \cdot_i, x_{i+1}, \ldots x_n)$.

If probabilities are taken into account we could consider the importance measure

$$I(i) = P\{0 = \phi(0_i, \underline{X}) < \phi(1_i, \underline{X}) = 1\}.$$

*Example 3.* For the previous system if $P\{X_1 = 1\} = .99$, $P\{X_2 = 1\} = .98$, $P\{X_3 = 1\} = .97$, then

$$I(1) = (.02)(.97) + (.98)(.03) = .0488$$

$$I(2) = (.01)(.97) + (.99)(.03) = .0394$$

$$I(3) = (.01)(.98) + (.99)(.02) = .0296$$

and so component 1 is most important and component 3 is least important as expected.

## 2. MULTISTATE MONOTONE SYSTEMS—FINITE CASE

If a system and/or its components have more than two modes of operation the previous type of structure function cannot adequately describe such a system. For example, if besides "on" and "off" states, there are modes of partial operation (or partial failure) a more complex mathematical description is required. Some basic groundwork has been laid by Barlow and Wu (1978) and by El-Neweihi, Proschan and Sethuraman (1978). In these studies they consider situations where component and system operation can vary from complete failure to perfect operation and possibly operate in a finite number of intermediate states.

The monotone structure functions defined by El-Neweihi et al. (1978) satisfy the assumptions:

(1) the system's operation (0 for complete failure, $M$ for flawless operation, and $1, 2, \ldots, M - 1$ for states of partial operation) is given by the function $\Phi(x_1, x_2, \ldots, x_n)$ where $x_i$ describes the operation of the $i^{th}$ component (each $x_i$ takes a value from 0 to $M$);

(2) $\Phi$ is nondecreasing which means that $\underline{x} \leqslant \underline{y}$ implies $\Phi(\underline{x}) \leqslant \Phi(\underline{y})$ (i.e., improving components improves the system);

(3) the system works no better than its best component and no worse than its worse component, i.e., $\min x_i \leqslant \Phi(\underline{x}) \leqslant \max x_i$.

Actually the Barlow-Wu model is a special case of the above and the El-Neweihi et al. model assumes a fourth relevance condition which specifies essentially that all components are relevant at all levels of operation. Griffith (1980) has examined similar models with variants of this relevance condition.

Using the above model and making various relevance assumptions on the components, El-Neweihi et al. (1978) and Griffith (1980) have shown that the principle (1) still holds for multistate monotone structure functions. Furthermore, Barlow and Wu (1978) and Griffith (1980) have defined concepts of multistate component importance. The importance concept of Griffith may be used to evaluate the effect of an improvement in component reliability on system reliability. This notion thus generalizes the result in the binary case.

Bounds on the probabilistic performance function of a multistate system have not been extensively discussed although El-Neweihi et al. (1978) have briefly addressed this problem. In order to consider the problem of computing bounds more fully, Block and Savits (1981a) have given a decomposition of the structure function of a multistate system in terms of binary structure functions. Thus using results about the binary functions new results including the computation of bounds can be obtained for the multistate structure functions.

The decomposition result is that for the multistate monotone function $\Phi$,

$$\Phi(\underline{x}) = \sum_{k=1}^{M} \phi_k(\underline{\alpha}(\underline{x}))$$

where $\phi_k(\underline{\alpha}(\underline{x})) = \max\limits_{\underline{y} \in U_k} \min\limits_{1 \leqslant i \leqslant n} \alpha_{i,y_i}(\underline{x})$, $\alpha_{i,y_i}(\underline{x}) = 1$ if $y_i \leqslant x_i$ and 0 otherwise and $U_k = \{\underline{y}\colon \Phi(\underline{y}) \geqslant k \text{ and } \underline{z} \underset{+}{<} \underline{y} \text{ implies } \Phi(\underline{z}) < k\}$. Here the $\phi_k$ is a binary function of binary variables, i.e., it is like the monotone function discussed previously. Consequently many results are known and can be applied to obtain results concerning the multistate function $\Phi$. In particular, bounds on $E(\Phi(\underline{X}))$, where $\underline{X}$ is a random state vector, have been obtained by Block and Savits (1981a). We let $\bar{P}_i(j) = P\{X_i > j\}$ which is the probability that the $i^{\text{th}}$ component is in a state higher than $j$ and assume that the components are independent. Defining

$$L_k = \{\underline{y}\colon \Phi(\underline{y}) \leqslant k \text{ and } \underline{y} \underset{+}{<} \underline{z} \text{ implies } \Phi(\underline{z}) > k\}$$

for $k = 0, 1, \ldots, M-1$ we have for example

$$\sum_{k=1}^{M} \coprod_{\underline{y} \in L_{k-1}} \prod_{y_i < M} \bar{P}_i(y_i) \leqslant E(\Phi(\underline{X})) \leqslant \sum_{k=1}^{M} \coprod_{\underline{y} \in U_k} \prod_{y_i > 0} \bar{P}_i(y_i - 1)$$

where $\prod\limits_{i=1}^{n} p_i = p_1 \cdot p_2 \cdot \cdots \cdot p_n$ and $\coprod\limits_{i=1}^{n} p_i = 1 - \prod\limits_{i=1}^{n} (1 - p_i)$.

Another application of the decomposition result is a characterization of those multistate structure functions which are of the Barlow and Wu (1978) type. A multistate structure function $\Phi$ is said to be of the Barlow and Wu type if there is some binary monotone structure function (with corresponding min path sets $P_1, \ldots, P_p$) such that

$$\Phi(\underline{x}) = \max_{1 \leqslant j \leqslant p} \min_{i \in P_j} x_i.$$

## 3. MULTISTATE MONOTONE SYSTEMS—CONTINUOUS CASE

The previous results were obtained for the case where systems and components can take the values 0, 1, 2, ..., $M$. If instead, systems and components can take a continuum of values, the previous results do not apply. Block and Savits (1981b) have obtained a decomposition result in this situation. For purposes of exposition we shall assume that the system and components can take any nonnegative value. Again we want to consider systems such that improvement in components leads to system improvement. Thus we consider a nondecreasing function $\Phi(x_1, x_2, \ldots, x_n)$ where each $x_i \geqslant 0$ for each of the $n$ components. For simplicity we also assume that $\Phi$ is continuous.

Here the sets defined by

$$U_t = \{\underline{x} | \Phi(\underline{x}) \geqslant t\} \text{ for } t \geqslant 0$$

are crucial. The decomposition for $\Phi(\underline{x})$ is now given by

$$\Phi(\underline{x}) = \int_0^\infty \phi(\underline{x}, t)\, dt$$

where $\phi(\underline{x}, t) = 1$ if $\underline{x} \in U_t$ and 0 otherwise. Furthermore, we can represent $\phi$ as

$$\phi(\underline{x}, t) = \max_{\underline{y} \in P_t} \min_{1 \leqslant i \leqslant n} \alpha_{i,y_i}(\underline{x})$$

where $P_t = \{\underline{y} | \Phi(\underline{y}) \geqslant t \text{ and } \underline{x} \underset{+}{\leqslant} \underline{y} \text{ implies } \Phi(\underline{x}) < t\}$ and $\alpha_{i,y_i}(\underline{x}) = 1$ if $y_i \leqslant x_i$ and 0 otherwise. In order to obtain a "min cut" representation for $\Phi$ it is necessary to restrict its domain of definition. This situation introduces some technical difficulties which are dealt with in detail in Block and Savits (1981b).

As in the finite state case, one use of the decomposition result is to obtain bounds on the system performance function $E(\Phi(\underline{X}))$. These results have been derived in Block and Savits (1981b). Since they are similar to the results in the finite state case, but are notationally more involved we omit them here.

Another use of the decomposition result is to obtain a characterization of those continuous multistate structure functions satisfying

$$\Phi(t\underline{x}) = t\,\Phi(\underline{x}) \text{ for all } t > 0.$$

(Functions satisfying the above are called homogeneous functions.) It is shown that we can represent such $\Phi$ as

$$\Phi(\underline{x}) = \max_{\underline{y} \in P_1} \min_{y_i \neq 0} y_i^{-1} x_i$$

where $P_1 = \{\underline{y}: \Phi(\underline{y}) \geqslant 1$ and for $\underline{x} \lneqq \underline{y}$, $\Phi(\underline{x}) < 1\}$. Hence all homogeneous multistate structure functions are essentially of the Barlow-Wu type.

## REFERENCES


Barlow, R.E. and F. Proschan (1975). *Statistical Theory of Reliability and Life Testing.* Holt, Rinehart and Winston, New York.

Barlow, R.E. and A.S. Wu (1978). Coherent systems with multistate components. *Math. of Operations Research*, *3*, 275-281.

Block, H.W. and T.H. Savits (1981a). A decomposition for multistate monotone systems. *J. Appl. Probab.*, to appear.

Block, H.W. and T.H. Savits (1981b). Continuous state monotone systems.

El-Neweihi, E., F. Proschan, and J. Sethuraman (1978). Multistate coherent systems. *J. Appl. Prob.*, 15, 675-688.

Griffith, W.S. (1980). Multistate reliability models. *J. Appl. Prob., 17, 735-744.*

# Recent Advances in Statistical Methods for System Reliability Using Bernoulli Sampling of Components

Bernard Harris* and Andrew P. Soms†

*University of Wisconsin*

**Abstract**

The paper describes briefly the history of statistical inference in systems reliability using Bernoulli sampling of components. Particular attention is given to the notion of Buehler optimality and its role in such problems. The authors discuss the role of Sudakov's inequality in obtaining bounds on Buehler optimal confidence limits and show that the Lindstrom-Madden method yields a Sudakov limit for series systems when a particular choice of ordering function is employed. Bounds for Buehler optimal confidence limits for parallel systems are obtained using the notion of Schur concavity.

## 1. INTRODUCTION

In this paper we examine the history of statistical inference in systems reliability using Bernoulli sampling of components with particular emphasis on the notion of Buehler optimality and its role in such problems. In particular, we focus on recent results of the authors which facilitate obtaining Buehler optimal bounds on the reliability of series and parallel systems. For series systems, these results employ a generalization of an inequality of Sudakov and for parallel systems, Buehler optimal bounds are obtained using the notion of Schur concavity when the number of failures is small. Estimates of the optimal

*University of Wisconsin-Madison

†University of Wisconsin-Milwaukee

Research supported by the Office of Naval Research under Contract No. N00014-79-C-0321 and the United States Army under Contract No. DAAG29-80-C-0041.

bounds for parallel systems are obtained for those cases in which the technique employing Schur concavity fails.

We suppose that the system has $k$ components and let $p_i$, $i = 1, 2, \ldots, k$ be the probability that the $i^{\text{th}}$ component does not fail. We further assume that the components are stochastically independent. Let $h(p_1, p_2, \ldots, p_k)$ be the probability that the system does not fail. Then the problem under discussion may be described as follows. The experimenter takes $N_i$ observations on each of the $k$ components and records $y_i$, the number of failures observed on the $i^{\text{th}}$ component. Then given this data, we wish to obtain a $1 - \alpha$ lower confidence limit on $h(p_1, p_2, \ldots, p_k)$.

Throughout the discussion, we restrict our attention to coherent systems. These may be described as systems in which the system fails when all components fail and the system functions when all components function. In addition, replacing a defective component by a good component will not cause a functioning system to fail. The reader is referred to R. E. Barlow and F. Proschan [1] for relevant details concerning the reliability function $h(p_1, \ldots, p_k)$ and properties of coherent systems.

The problem discussed herein arises naturally in many situations relevant to acquisition decisions. For such an illustration consider a "one-shot device," namely, one which is to perform a specific function at a specific time, so that the lifetime of the device is not a consideration. Then, you are interested in assuring that the probability that the device will function properly when it is to be used is at least a specified number.

For a second application, consider a system which is to function for a specified length of time, known as the mission time. If the only data available is the number of failures of each component before the mission time, then the model of this paper is appropriate.

Throughout the paper, we emphasize the specific cases of series and parallel systems.

## 2. BUEHLER OPTIMAL CONFIDENCE LIMITS FOR SYSTEM RELIABILITY

We now introduce the notion of Buehler optimality with respect to the ordering function $g(\tilde{x})$, where $\tilde{x} = (x_1, x_2, \ldots, x_k)$, $x_i = N_i - y_i$, $i = 1, 2, \ldots, k$. In Buehler (1957), the following principle was proposed and applied to obtaining confidence limits for the product of two binomial parameters. This corresponds to proposing an optimal solution to the problem of determining lower confidence limits for the reliability of series systems of two components.

$g(\tilde{x})$ is an ordering function if for $\tilde{x}^{(1)} = (x_{11}, x_{12}, \ldots, x_{1k})$ and $\tilde{x}^{(2)} = (x_{21}, x_{22}, \ldots, x_{2k})$ with $x_{1i} \leqslant x_{2i}$, $i = 1, 2, \ldots, k$, we have $g(\tilde{x}^{(1)}) \leqslant g(\tilde{x}^{(2)})$. In the original formulation, one orders the $\prod_{i=1}^{k} (N_i + 1)$ sample outcomes and lists the corresponding values of $g(\tilde{x})$. Then calculate

$$P_{\tilde{p}}\{g(\tilde{X}) \geqslant u\}. \tag{1}$$

This is the probability, for fixed $\tilde{p}$, of getting an outcome as good as or better than $u$ since $g(\tilde{x})$ is ordered consistently with the number of functioning components observed in the sample.

If $\tilde{x}_0$ is the observed sample outcome,

$$Q_\alpha(\tilde{x}_0) = \{\tilde{p}: P_{\tilde{p}}(g(\tilde{X}) \geqslant g(\tilde{x}_0)\} \geqslant \alpha, \tag{2}$$

then $Q_\alpha$ is a $1 - \alpha$ confidence region for $\tilde{p}$ determined by $g(\tilde{x})$. This can be extended to a $1 - \alpha$ lower confidence bound for $h(\tilde{p})$ by letting

$$\underline{h}(\tilde{u}) = \inf\{h(\tilde{p}): P_{\tilde{p}}(g(\tilde{X}) \geqslant u) \geqslant \alpha\}, \tag{3}$$

for $0 < \alpha < 1$. It can be shown that if $\underline{l}(u)$ is any other system of lower confidence limits based on the ordering function $g(\tilde{x})$, then $\underline{l}(u) \leqslant \underline{h}(u)$ for all $u$. Further details on this optimality property and a proof of the above assertion may be found in Harris and Soms [8].

Unfortunately, while this does give a procedure for obtaining best lower confidence bounds once $g(\tilde{x})$ has been specified, this does not say how to choose $g(\tilde{x})$. For various kinds of systems, many researchers have proposed different ordering functions and quite a few statisticians have suggested procedures which have the Buehler optimality property asymptotically but not necessarily for any fixed sample size. In addition the literature contains many suggested procedures which are not optimal but possess some other favorable attribute such as ease of computation, while not deviating too far from optimality. As originally described by Buehler many procedures employing Buehler optimality are virtually incomputable for $k > 2$.

Other properties which some writers have considered desirable are exactness and conservatism. Specifically, a confidence interval procedure is exact if there are parameter points such that for those points, $\geqslant \alpha$ in (3) can be replaced by equality. In the case of (3), this is true for a large class of ordering functions, since $h(\tilde{p})$ is continuous and the infimum will be attained. A confidence interval procedure is conservative if the true confidence coefficient is at least $1 - \alpha$ for all parameter points. Clearly, many asymptotic approximations and other approximate procedures will violate this.

We discuss this by looking at some specific types of systems, beginning with series systems.

## 3. LOWER CONFIDENCE LIMITS FOR THE RELIABILITY OF SERIES SYSTEMS

A natural first attempt in selecting an ordering function for series systems is to choose a point estimator for $h(\tilde{p}) = \prod_{i=1}^{k} p_i$. Such an estimator is the maximum likelihood estimator given by

$$h(\hat{p}) = \prod_{i=1}^{k} \frac{X_i}{N_i}. \tag{4}$$

In the situation where the $p_i$'s are close to unity, the high reliability case, we can write

$$h(\hat{p}) = \prod_{i=1}^{k} \frac{N_i - Y_i}{N_i} \tag{5}$$

and if the $N_i$'s are large, this is commonly approximated by

$$h(\hat{p}) = \prod_{i=1}^{k} (1 - \frac{Y_i}{N_i}) \sim 1 - \sum_{i=1}^{k} \frac{Y_i}{N_i}. \tag{6}$$

Then, this suggests replacing the original assumption of the binomial distribution by the Poisson distribution and this alternative has been exploited by many practitioners. In particular, $1 - h(\hat{p})$ is regarded as being a linear combination of Poisson random variables with expected value $\Sigma\lambda_i/N_i$.

Some alternative ordering functions that have been considered are

$$\sum a_i Y_i + z_\alpha(\Sigma a_i^2 y_i)^{1/2}, \tag{7}$$

where $z_\alpha$ satisfies $1 - \phi(z_\alpha) = \alpha$, $\Phi(x)$ is the standard normal distribution function, $a_1 \geqslant \cdots \geqslant a_n$ and $a_i = (n_i \Sigma \frac{1}{n_i})^{-1}$. This choice was proposed by M. V. Johns, Jr. [10] and is somewhat suggestive of (6) in that it employs a weighted normal deviate suggested by the Poisson approximation noted previously.

Buehler [2] ordered by computing separate $(1 - \alpha)^{1/k}$ confidence bounds. The product of the separate confidence limits is the final lower confidence limit.

It should be noted that Buehler directed most of his discussion to parallel systems and all of his approximations and numerical examples are appropriate only for that application. However, using the duality noted later, the general discussion in his paper can be applied equally well to both series and parallel systems.

Closely related to the maximum likelihood estimator and sometimes employed as an ordering function are estimators of the form

$$g(\tilde{x}) = \prod_{i=1}^{k} (X_i - \alpha_i)/(N_i - \beta_i), \tag{8}$$

where $\alpha_i$, $\beta_i$ may be dependent on the $X_i$'s and $N_i$'s. Naturally, $\alpha_i$ and $\beta_i$ must satisfy conditions such that $g(\tilde{x})$ will be asymptotically equivalent to the maximum likelihood estimator (5).

An example of this approach is given by A. Madansky [13], who calculated the Wilks' likelihood ratio $L(\tilde{x})$ and used the asymptotic result that $-2 \log L(\tilde{x})$ has asymptotically a chi-square distribution with one degree of freedom to obtain a lower confidence limit to the reliability function $h(\hat{p})$. He compared the results obtained using the likelihood ratio to those that would be obtained employing the maximum likelihood estimator, where the distribution of the maximum likelihood estimator is approximated by employing asymptotic normality and its usual asymptotic variance for the maximum likelihood estimator. Madansky refers to this as the linearization method.

For this case, the approximate lower confidence limit has the form

$$\prod_{i=1}^{k} \frac{X_i}{N_i} - (q(\tilde{X}))^{1/2} z_\alpha, \tag{9}$$

where

$$q(\tilde{X}) = \sum_i \left(\frac{\partial h(\hat{p})}{\partial \hat{p}_i}\right)^2 \hat{p}_i(1 - \hat{p}_i). \tag{10}$$

For both of these techniques, the convergence to the asymptotic distribution is not uniform and substantially larger sample sizes are required near the boundary of the parameter space in order for the limiting distribution to provide a satisfactory approximation.

Some technical improvements using the likelihood ratio technique are given in C. Mack [12], but their relationship with the likelihood ratio procedure is not identified in that paper.

The Lindstrom-Madden method, as described in Lloyd and Lipow [11] is of fundamental importance to this discussion. Let $h(\hat{p})$ denote the maximum likelihood estimator as in (5) and let $\underline{N} = \min_{1 \leqslant i \leqslant k} N_i$. Then regard $\underline{N}(1 - h(\hat{p})) = V$ as the number of failures in $\underline{N}$ Bernoulli trial and use the usual method of obtaining a lower confidence limit for a single binomial proportion. In general, $V$ will not be an integer, so that the interpretation as a binomial random variable is not completely justified. One proceeds by interpolation in either the tables of the binomial distribution or in the tables of the incomplete beta function. Some forms of nonlinear interpolation have also been proposed and investigated.

Although this procedure is widely used by engineers, it is our impression that statisticians tended to view it with some skepticism. However, extensive simulations and numerical computations made prior to the issuance of the Handbook for the Calculation of Lower Statistical Confidence Bounds on System Reliability [3] suggested that it performs as well as any of the competing methods under investigation in the region of high reliability. In addition the simulations strongly suggested that it met the requirement of being conservative, that is, that the true confidence level is always at least as high as the nominal confidence level. A proof of this fact using an inequality due to Sudakov [21] is given in Harris and Soms [8]. The Lindstrom-Madden technique has been recommended for series systems in the above mentioned handbook.

Closely related to the Lindstrom-Madden method is the method of "key test results" introduced by K. A. Weaver [22] and extended by A. Winterbottom [23]. The maximum likelihood estimate $h(\hat{p})$ is calculated. Let $\underline{N} = \min_i N_i$. Calculate $[\underline{N}h(\hat{p})]$. This of course is always an integer. Then find the usual binomial confidence limit for $N$ and $[Nh(\hat{p})] = x$. If $Nh(\hat{p})$ is an integer, then the Lindstrom-Madden method and the key test method coincide. Since we have shown [8] that the Lindstrom-Madden method is conservative, it follows that the method of key test results is at least as conservative.

Easterling's [4] modified maximum likelihood method also employs the maximum likelihood estimator. From the usual asymptotic theory for maximum likelihood estimation, the estimator $h(\hat{p})$ has an asymptotic variance given by

$$\sigma^2_{h(\hat{p})} = \sum_{i=1}^{k} \left( \frac{\partial h(p)}{\partial p_i} \right)^2 \mathrm{Var}\,(\hat{p}_i). \tag{11}$$

Replace $p_i$ by $\hat{p}_i$ in the above, thereby obtaining (10), and set

$$\hat{\sigma}^2_{h(\hat{p})} = \frac{h(\hat{p})\,(1 - h(\hat{p}))}{\hat{n}}. \tag{12}$$

Then regard $\hat{n}$ as the binomial sample size with $\hat{n}h(\hat{p})$ successes observed and obtain the lower confidence interval for a single binomial parameter, interpolating as in the Lindstrom-Madden method.

A further modification is suggested in which $\hat{n}$ and $\hat{x}$ are replaced by the next integer. This is designated as the MMLI method. The suggestion that this may be satisfactory is deduced from an examination of the deficiencies noted in the linearization (maximum likelihood) method by Madansky.

In a related investigation, J. L. Epstein [5] considered the problem of confidence sets for the product of two binomial parameters. This can be interpreted as either a parallel or a series system in the reliability context. Epstein was motivated by some biomedical applications in which the assumption of high reliability for each of the components is not as natural as it is for the engineering reliability context.

In studying this question, Epstein considered the two ordering functions

$$g_1(\tilde{x}) = \frac{X_1 X_2}{N_1 N_2}, \quad g_2(\tilde{x}) = \frac{(X_1 + 1)\,(X_2 + 1)}{N_1 N_2}.$$

He concluded that the second was preferable to the first, since the partition of the sample space induced by the second is finer than that induced by the first. This effect is particularly pronounced when either $X_1$ or $X_2$ is zero. This is not, however, a significant factor for series systems with high reliability. It is, nevertheless, quite important for parallel systems with high reliability, as we shall subsequently observe.

Apparently ideas similar to those deduced by Epstein motivated the development of the technique suggested by Sandia Corporation and referred to as CONLIM [6].

Let $h(\tilde{p})$ be the reliability function of any coherent system. Then let

$$g(\tilde{x}) = h(\hat{p}), \tag{13}$$

where $\hat{p} = (\hat{p}_1, \ldots, \hat{p}_k)$ and

$$\hat{p}_i = (X_i + 1)/(N_i + 2), \; 1 \leqslant i \leqslant k. \tag{14}$$

For the series system, this reduces to

$$g(\tilde{X}) = \prod_{t=1}^{k} (X_i + 1)/(N_i + 2). \tag{15}$$

The particular choice of $\hat{p}_i$ leads one to suspect that this was motivated by Bayesian considerations employing independent uniform priors on $0 \leqslant p_i \leqslant 1$. The computer program which uses (13) to calculate lower confidence bounds for system reliability is capable of dealing with a large variety of systems, but unfortunately a substantial amount of computer time is often required in order to calculate the lower confidence limit.

The Poisson approximation methods, alluded to earlier, have principally been exploited by statisticians from the Soviet Union.

In particular, Pavlov [19] used the ordering function (6), by defining the parameter for which the confidence limit is sought as

$$\rho = \frac{\lambda_1}{N_1} + \frac{\lambda_2}{N_2} + \cdots + \frac{\lambda_k}{N_K}. \tag{16}$$

Then, choose $M$ so that $MN_1, \ldots, MN_k$ are all approximately integers. This results in

$$M\rho = M_1\lambda_1 + M_2\lambda_2 + \cdots + M_k\lambda_k \tag{17}$$

and $\Sigma\, M_i Y_i$ is a suitable statistic for estimation of the parametric function (17) and therefore can be employed to get confidence bounds on $\rho$.

As a specific illustration, consider $k = 3$, $N_1 = 600$, $N_2 = 300$, $N_3 = 100$. Then, from (6),

$$h(\tilde{p}) \sim 1 - \frac{q_1}{N_1} - \frac{q_2}{N_2} - \frac{q_3}{N_3}$$

and

$$600(1 - h(\tilde{p})) \sim 600\left(\frac{\lambda_1}{600} + \frac{\lambda_2}{300} + \frac{\lambda_3}{100}\right),$$

so that $Y_1 + 2Y_2 + 6Y_3$ is an appropriate statistic to use for estimation of $h(\tilde{p})$ and for other inferential purposes.

Another method for employing Poisson approximations has been given by Mirniy and Solov'yev [18]. This technique may be described as follows. Let

$$h(\tilde{p}) = \prod_{i=1}^{k} p_i = e^{\sum_{i=1}^{k} \log p_i}. \tag{18}$$

For $p_i$ near one, as is appropriate for high reliability,

$$\log p_i \sim -(1 - p_i).$$

Let $\lambda_i = N_i(1 - p_i)$, $i = 1, 2, \ldots, k$. Then construct an upper confidence limit for $\sum_{i=1}^{n} \lambda_i$ and then an approximate upper confidence limit for

$$e^{-\sum_{i=1}^{k}(1-p_i)} \sim e^{-\sum_{i=1}^{k}\lambda_i/N_i} \tag{19}$$

is given by $e^{\max(-\Sigma\lambda_i/N_i)}$, where the maximum is over the set $\{\sum_{i=1}^{k} \lambda_i \leqslant \Delta\}$, $\Delta$ the upper confidence limit referred to above. This procedure is completely analogous to the procedure using (2) and (3) except that an upper confidence bound is being constructed for the unreliability.

This brief survey makes no pretense of being complete. In particular, the reader is referred to the paper by Joan Rosenblatt [20], which contains a survey of many of the early techniques employed for this problem. Likewise, the book by N. Mann, R. E. Schafer, and N. D. Singpurwalla [14] should be noted. In particular Chapter 10 is relevant to this discussion and a commonly employed method, called the AO method is described therein. Also Bayesian methods, such as these proposed by D. Mastran [16] and D. Mastran and N. D. Singpurwalla [17], have been omitted, since these are obtained using a different set of principles than the methods which are compared in this paper.

Given this large number of available techniques, how does one make a selection? To answer this question, in the course of the preparation of the Handbook for the Calculation of Lower Statistical Confidence Bounds on System Reliability [3], an extensive investigation of various procedures was carried out. This investigation employed both simulation methods and numerical calculations. Among the requirements sought for was conservatism of procedures. However, while procedures that were found to be optimistic (nonconservative) were regarded as not satisfactory, attention was paid as well to accuracy, namely, that the true confidence coefficient should be as close to $(1 - \alpha)$, the nominal confidence coefficient, as possible. Ease of computation was also considered. Finally, performance of the statistical procedures for high reliability was regarded as being of much more significance than performance at low reliabilities.

This investigation established among other things, that the AO method and Easterling's two proposals are nonconservative. CONLIM satisfied every requirement with the exception of ease of computation.

A surprising conclusion of this investigation was that the Lindstrom-Madden method was found to be conservative, but not excessively conservative. In short, it was the opinion of the committee entrusted with the preparation of this handbook, that the Lindstrom-Madden method should be adopted for use and be the method recommended in the handbook. Since the investigation undertaken in preparation of the handbook employed simulation and the computation of a large number of numerical examples, a theoretical explanation for the impressive performance of the Lindstrom-Madden method is needed.

The answer to this is provided in Harris and Soms [8] and depends on the application of an inequality of Sudakov [21]. The description of these results follow.

Let

$$I_p(r,s) = \frac{1}{B(r,s)} \int_0^p t^{r-1}(1-t)^{s-1}dt.$$

Then, it is well known that

$$\sum_{i=0}^{y} \binom{n}{i} p^{n-i} q^i = I_p(n-y,\ y+1). \tag{20}$$

Let $u(n,\ y,\ \alpha)$ satisfy

$$\alpha = I_{u(n,y,\alpha)}(n-y,\ y+1) \tag{21}$$

and assume that $N_1 \leqslant N_2 \leqslant \cdots \leqslant N_k$. Let

$$g(\tilde{X}) = \prod_{i=1}^{k} (\tilde{X}_i / N_i),$$

and let

$$y_1 = N_1 q_0 = N_1 (1 - \prod_{i=1}^{k} \frac{x_{0i}}{N_i}), \tag{22}$$

where $x_{0i}$ is the observed value of $X_i$. Then Sudakov showed that

$$u(N_1, y_1, \alpha) \leqslant b \leqslant u(N_1, [y_1], \alpha), \tag{23}$$

where $b$ is the $1-\alpha$ Buehler optimal lower confidence limit for $h(\tilde{p})$. Now $u(N_1, y_1, \alpha)$ is the Lindstrom-Madden solution. Thus, it follows that if $y$ is an integer, the Lindstrom-Madden solution is optimal.

In Harris and Soms, a generalization of the Sudakov inequality was employed to show that (23) holds when $y_1$ is computed for any of a large class of ordering functions in addition to the maximum likelihood estimator. This class includes

$$g(\tilde{X}) = 1 - \sum_{i=1}^{k} \frac{Y_i}{N_i},$$

$$g(\tilde{X}) = \prod (X_i + \alpha_i),\ \alpha_i \geqslant 0,$$

where $N_{i+1}\alpha_i \geqslant \alpha_{i+1} N_i$, and for $k = 2$ the choice (7) employed by Johns. In addition, an improvement on the upper bound in (23) was obtained. In short, if

$$y_i^* = \max \{y_i : g(\tilde{\tilde{y}}) \leqslant g(\tilde{y}_0),\ y_i = N_i - x_i,\ y_{0i} = N_i - x_{0i}\} \tag{24}$$

and $\tilde{\tilde{y}} = (0, \ldots, 0, y_i, 0, \ldots, 0)$, $y_1^* = y_1$, then

$$u(N_1, y_1, \alpha) \leqslant b \leqslant \min_i u(N_i, [y_i^*], \alpha), \tag{25}$$

holds for a class of ordering functions, which includes many commonly in use. For the relevant details, see Harris and Soms. The left-hand side of (25) may be regarded as a generalization of the Lindstrom-Madden method.

## 4. LOWER CONFIDENCE UNITS FOR THE RELIABILITY OF PARALLEL SYSTEMS

To some extent, there is a duality between series and parallel systems. For a parallel system,

$$1 - h(\tilde{p}) = \prod_{i=1}^{k} q_i = \prod_{i=1}^{k} (1 - p_i). \tag{26}$$

Thus, interchanging reliability and unreliability, success and failure, $p_i$ and $1 - p_i = q_i$ converts any probability statement about a series system into an "equivalent" statement about a parallel system. Unfortunately, this duality will not extend as readily to the statistical inference aspects. The reasons are that the above duality will result in conservative confidence bounds being changed into non-conservative bounds. Also high reliability becomes high unreliability, an area of little interest to practitioners. Further, the Poisson approximation techniques employed in the study of one type of system will not carry over to the other, since small failure probabilities become small success probabilities, again a situation of virtually no interest.

Nevertheless, some parts of the previous discussion for series systems do carry over to parallel systems.

For this reason, it is convenient to replace the problem of finding a lower confidence limit for the reliability by the problem of finding an upper confidence level for the unreliability. Then, to make the analogy with series systems clearer, we will denote the unreliability by $\bar{h}(\tilde{p})$ and the ordering function based on failures by $\bar{g}(\bar{X})$, where $\bar{X}$ denote failures, that is

$$1 - h(\tilde{p}) = \bar{h}(\tilde{p}),\ N_i - X_i = \bar{X}_i,$$

and $\bar{g}(\bar{X})$ will often be an estimator of the unreliability.

Buehler's [2] method of computing separate $(1 - \alpha)^{1/k}$ confidence bounds carries over, however, the Poisson approximations that he uses to get specific numerical values are valid only for parallel systems.

Madansky's [13] method using the Wilk's likelihood ratio statistic is equally valid for parallel systems, since in employing a continuous approximation, namely the chi-square distribution, the transference from conservatism to optimism does not take place. Thus, all comments about the use of the technique for series systems apply essentially unchanged for parallel systems. Similarly, the comments about use of the asymptotic theory for the maximum likelihood estimator remain unchanged.

The observations of Epstein [5] that the ordering functions $(\bar{X}_1 + 1)(\bar{X}_2 + 1)/N_1N_2$ is preferable to $\bar{X}_1\bar{X}_2/N_1N_2$ is far more relevant to parallel systems than to series systems. To see this, note that the maximum likelihood estimator of $\prod_{i=1}^{k} q_i$ is $\prod_{i=1}^{k} \bar{X}_i/\prod_{i=1}^{k} N_i$, where $X_i$ is the number of observed failures of the $i^{th}$ component. In the case of high reliability, $\bar{X}_i$ will tend to be small and one would like to believe that it will frequently be zero. The difficulty with an ordering function such as $\prod_{i=1}^{k} (\bar{X}_i/N_i)$ is that for $k = 2$, the sample outcomes $\bar{X}_1 = 0$, $\bar{X}_2 = 0$ is not distinguished from $\bar{X}_1 = 0$, $\bar{X}_2 = 100$, when $N_1 = N_2 = 100$. More precisely, the partition induced by $\bar{g}(\bar{X})$ is then very coarse when at least one $\bar{X}_i$ is zero. Consequently, the use of a finer partition for the high reliability situation may produce less conservatism and greater accuracy. Presumably, similar considerations motivated the choice of $\hat{p}_i$ for the CONLIM method, which in this case gives

$$\bar{g}(\bar{X}) = \prod_{i=1}^{k} (\bar{X}_i + 1)/(N_i + 2). \tag{27}$$

A somewhat different technique for obtaining an upper confidence limit on $\bar{h}(\tilde{p})$ is given in Harris [7], where the exponential family theory is used to obtain the uniformly most accurate unbiased upper confidence limit. However, this technique does not have a simple description in terms of an ordering function.

The specific question of Buehler optimality for parallel systems was considered in Harris and Soms [9]. There the following results were obtained.

Let

$$\bar{g}(\bar{X}) = \prod_{i=1}^{k} (\bar{X}_i + d), \ 1 < d < 1.5. \tag{28}$$

Then replacing the binomial distribution by the Poisson approximation, let $\tilde{z}$ be any failure vector with $z_i \leqslant 5$ for some $1 \leqslant i \leqslant k$ and $z_i = 0$ otherwise, and let

$$A_{\tilde{z}} = \{\tilde{x} : \bar{g}(\tilde{x}) \leqslant \bar{g}(\tilde{z})\}. \tag{29}$$

Then for any $\tilde{x} \in A_{\tilde{z}}$,

$$P_{\tilde{\lambda}} \{\bar{g}(\bar{X}) \leqslant \bar{g}(\tilde{x})\}$$

is a Schur-concave function of $-\ln \lambda_i$, $i = 1, 2, \ldots, k$, where $\tilde{\lambda}$ is the parameter vector for the Poisson approximation obtained by letting $\lambda_i = N_i\bar{q}_i$, $i = 1, 2, \ldots, k$. The reader is referred to A. W. Marshall and I. Olkin [15] for details on the properties of Schur-concave functions. This enables one to conclude that for $\tilde{x}_0 \in A_z$ the Buehler optimal upper confidence limit for $\prod_{i=1}^{k} \lambda_i$ is obtained from the solution of

$$P_{\tilde{\lambda}}{}^*\{\bar{g}(\bar{X}) \leqslant \bar{g}(\tilde{x}_0)\} = \alpha, \tag{30}$$

where $\tilde{\lambda}^* = (a, \ldots, a)$, namely has all components identical. This Schur-concavity argument may be extended to other ordering functions than (28). In particular, the asymptotic form of the choice used by Buehler [2] also satisfies the necessary conditions.

Unfortunately, if the observed vector of failures lies outside $A_{\bar{z}}$, this conclusion does not hold for the ordering function (28). Here, upper and lower bounds for the Buehler optimal upper confidence limit have been constructed. The reader is referred to the original paper [9] for details.

## ADDENDUM – ADDED IN PROOF

The authors have recently discovered an error in the proof of one of the lemmas employed in the proof of Sudakov's inequality. This observation has led to the discovery of counterexamples for the inequality itself. Present evidence suggests strongly that this will not have any effect on applications, since the known counterexamples are all for confidence levels quite small. Extensive numerical evidence, including the work prior to the preparation of the Handbook [3] indicates that violations of this inequality are expected to occur only in this region. The authors are continuing to investigate the inequality and hope to have a definitive answer shortly.

## REFERENCES

R.E. Barlow and F. Proschan, *Statistical Theory of Reliability and Life Testing, Probability Models*, Holt, Rinehart and Winston, Inc., New York, 1975.

R.J. Buehler, "Confidence limits for the product of two binomial parameters," *J. Amer. Statist. Assoc.* 52 (1957), 482-493.

Defense Advanced Research Projects Agency, Handbook for the Calculation of Lower Statistical Confidence Bounds in System Reliability, 1980.

R.G. Easterling, "Approximate confidence limits for series reliability," *J. Amer. Statist. Assoc.* 67 (1972), 220-222.

J.E. Epstein, "Upper Confidence limits on the product of two binomial parameters," Cornell University, Master of Science Thesis, 1967.

R.D. Halbgewachs, R.C. Mueller and F.W. Müller, "Classical Upper Confidence Limits for the Failure Probability of Systems," Research Report SLA-73-0563, Sandia Laboratories, (1973).

B. Harris, "Hypothesis testing and confidence intervals for products and quotients of Poisson parameters with applications to reliability," *J. Amer. Statist.*, 66 (1971), 609-613.

B. Harris and A.P. Soms, "Bounds for Optimal Confidence Limits for Series Systems," University of Wisconsin, Statistics Department Technical Report #611 (1980).

B. Harris and A.P. Soms, "Optimal Upper Confidence Limits for Products of Poisson Parameters with Applications to the Interval Estimation of the Failure Probability of Parallel Systems," University of Wisconsin, Statistics Department Technical Report #613 (1980).

M.V. Johns, Jr., "Confidence Bounds for Highly Reliable Systems," Stanford University Technical Report, (1976).

D.K. Lloyd and M. Lipow, *Reliability: Management, Methods and Mathematics*, Prentice Hall, Englewood Cliffs, New Jersey, 1962.

C. Mack, "Confidence limits for the product of binomial parameters and related problems," *New J. Statistics and Operations Research*, 3 (1967), 3-13.

A. Madansky, "Approximate confidence limits for the reliability of series and parallel systems," *Technometrics*, 7 (1965), 495-503.

N.R. Mann, R.E. Schafer, and N.D. Singpurwalla, *Methods for Statistical Analysis of Reliability and Life Data*, John Wiley and Sons, New York, 1974.

A.W. Marshall and I. Olkin, *Inequalities: Theory of Majorization and Its Applications*, Academic Press, New York, 1979.

D. Mastran, "A Bayesian approach to assessing the reliability of Air Force reentry systems," Proc. of the ASME Reliability and Maintainability Symposium, Gordon and Breach, New York, 1969.

D. Mastran and N.D. Singpurwalla, "A Bayesian estimation of the reliability of coherent structures," *Operations Research*, 26 (1978), 663-672.

R.A. Mirniy and A.D. Solov'yev, "Estimation of the Reliability of a System from the Results of Tests of its Components," *Kibernetiky na Sluzhby Kommunizmy*, 2, Energiya, Moscow, (1964).

I.V. Pavlov, "A Confidence Estimate of System Reliability from Component Testing Results," *Izvestiya Akad. Nauk. Tech. Kibernetiky*, 3, 52-61 (1973).

J.R. Rosenblatt, "Confidence limits for the reliability of complex systems," in *Statistical Theory of Reliability*, M. Zelen, ed., University of Wisconsin Press, Madison (1963).

R.S. Sudakov, "On the Question of Interval Estimation of the Index of Reliability of a Sequential System," *Engineering Cybernetics*, 12, 55-63 (1974).

K.A. Weaver, "The determination of the lower confidence limit on the reliability of serial system," Proc. of the NATO Conference on Reliability Theory, Turin, 1969, Gordon and Breach, New York, 1971.

A. Winterbottom, "Lower Limits for Series System Reliability from Binomial Data," *Journal of the American Statistical Association*, 69, 782-8 (1974).

# Repairable Systems: Reliability's Stepchild

Harold Ascher

*Naval Research Laboratory*

Harry Feingold

*David W. Taylor Naval Ship Research and Development Center*

**Abstract**

Most real systems are designed to be repaired, rather than overhauled or replaced, after failure. In spite of this empirically obvious fact, the preponderance of the reliability literature treats nonrepairable items. Moreover, to the extent that repairable systems are considered at all, the emphasis is on (the necessarily complicated) probabilistic modeling of complex systems. This ironically, has contributed to the situation where even basic, easily understood concepts and simply implemented/interpreted *statistical* techniques for repairable systems are seldom considered in papers and, astoundingly, are virtually ignored by reliability texts.

We have elaborated the above points in a book to be published by Marcel Dekker. In this paper, therefore, we consider the area of greatest interest to researchers, namely, topics for future research.

## INTRODUCTION

A system can be defined, for the purposes of this paper, as a collection of two or more parts which is designed to perform one or more functions. A repairable system is a system which, after failing to perform one or more of its functions satisfactorily, can be restored to fully satisfactory performance by replacing, at most, some of its constituent parts. Since such a system can fail two or more times, the pattern of times between successive failures is of fundamental importance. For example, if these times are tending to become larger and larger, this is an indication that reliability growth is occurring and conversely, if these times are getting smaller, i.e., if the system is deteriorating, thought must be given to reversing this trend or discarding the system.

It is empirically obvious that most "real world" systems are designed to be repairable. It would appear, therefore, that there would be an extensive literature on probabilistic modeling and statistical analysis of such systems. However, with the partial exception of probabilistic cost models and reliability growth models, the entire area of repairable systems has been seriously neglected in the reliability literature. For example, when one of us remarked to a colleague that, to a first approximation, almost nobody in the reliability field understands anything about repairable systems, he replied that that assessment would hold as a second approximation as well! We have pointed out many of the misconceptions about repairable systems rampant in the reliability field in a series of papers, Ascher and Feingold (1969, 1978a, 1978b, 1979). We have also stressed the curious omission from the reliability texts —and to a large degree, the entire reliability literature—of even the basic concepts of repairable systems. Moreover, we have pointed out that even very simply implemented and interpreted *optimal* statistical tests have been ignored by reliability texts and given far too little emphasis in the rest of the literature. We are presently summarizing the state-of-art of the probabilistic modeling and statistical analysis of repairable systems in a monograph, Ascher and Feingold (1981). This monograph also discusses, in detail, the many misconceptions which are widely accepted in the reliability community and the reasons why these misconceptions persist. Rather than attempting to summarize all this material in this paper in necessarily abbreviated form, we will restrict our attention to recommendations for future research on repairable systems.

We want to emphasize that in addition to the need for research, there are many techniques presently available which can be profitably applied to the analysis of repairable systems. One of the main reasons these techniques are not being applied is that the *need* for such techniques—in distinction to techniques for nonrepairable items—is unknown to the reliability community. The problem can be summarized by the following vicious cycle: More will be said about this topic under the discussion of updating of MIL-STD-721, the standard covering reliability definitions. We note that Thompson (1981) has also discussed some of these same problems and their causes.

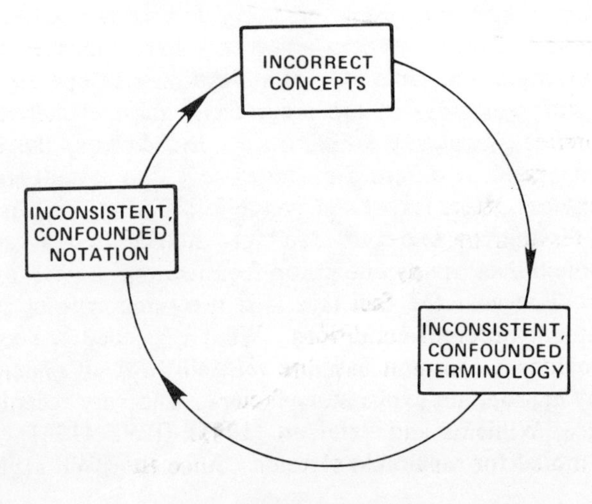


VISCIOUS CYCLE

## RESEARCH TOPICS

In this section we will discuss a number of problem areas which, in our opinion, have great potential for providing improved techniques for the probabilistic modeling and statistical analysis of repairable systems. In our last section, we will discuss the need for improved Military Standards and Handbooks on this subject.

a. Multivariate Regression Models

In the series of papers alluded to above, we have stressed the need for more applications of point process techniques to repairable systems. It must be recognized, however, that in many cases the point process approach is not practical because of inadequate sample sizes. That is, in many cases there will be no more than a few failures per system, at least over the available observation period. Under such circumstances, data for two or more—preferably many more—systems would have to be pooled, in order to use the point process approach. However, unless it is assumed that the data from each system are modeled by a homogeneous Poisson process (HPP) or nonhomogeneous Poisson process (NHPP), the probabilistic law for the pooled process is unknown. In some cases it will be reasonable to make the HPP or NHPP assumption. However, because this assumption cannot be checked against the data—because of the small sample sizes—alternative approaches are desirable in all cases where point process techniques cannot be applied to the data from each system.

In some situations there will be only a few systems available for analysis, each of which have failed no more than a few times. What this amounts to, of course, is a very small data base which is inadequate for statistical analysis. It is often the case, however, that there are a large number of systems available for analysis. In the past, when this has been true, the approach has been to assume that times to first failure of these systems are independent and identically distributed (IID). It has also been assumed that the data for each of the interarrival times are IID, not necessarily with the same distribution as time to first failure. What this type of approach has ignored is that if there are a large number of different systems, there probably are known differences in the installation/operation of these systems which may have a marked effect on their reliability. For example, the same type of system may be operated on different platforms (e.g., different ships of the same class, ships of different classes or even on say, surface ships and submarines). In addition, the same type of system may be operated in different positions (e.g., on a multiengine aircraft) with different ambient stress levels and possibly, different performance criteria. It often is not feasible to segregate data into homogeneous groups because there are not enough data in any one group for meaningful analysis. Moreover, such segregation disregards the fact that it is the same type of system that is being operated under different conditions. What is needed is a regression type model which provides a common baseline reliability for all systems of a given type, modified by appropriate explanatory factors. The very recently introduced model of Prentice, Williams and Peterson (1981) (PWP (1981)) provides the basis for such a model for repairable systems. Since the PWP (1981) technique has been developed in a biometry context, some modification will be required for optimum use in reliability problems. For example, the time scale usually used in biometry is the total calendar time since entry into a study; PWP (1981) considers this scale as well as calendar time since last failure (their term). In reliability the scale almost always used for repairable systems is the number of operating hours since the most recent repair since far too much emphasis is placed on the assumption of renewal by each repair. Usually, total operating time, regardless of the number of failures/repairs will be a more useful time scale than the backward recurrence time to the most recent failure. Additionally, other time scales will be useful, such as operating time since most recent overhaul, total calendar time since initial startup and possibly, backward recurrence time, as well. Implementing changes such as these will result in an extraordinarily powerful tool for the analysis of repairable systems.

Because of its obvious advantages, we should be able to say that the prognosis for adoption of this model by the reliability community is good. Based on the track record, however, we are not optimistic about such acceptance. First, *any* technique for repairable systems has had difficulty in being accepted. Secondly, the PWP (1981) model is an extension of Cox's (1972) proportional hazards model. Cox's technique has been very widely implemented in the biometry field. Potentially, it is equally applicable to nonrepairable items but we do not know of a single application to reliability problems! Obviously, the PWP (1981) model will not be implemented in the reliability field merely because of its merits.

### b. Superposition of Point Processes other than HPP's and NHPP's

When a finite number of point processes are superposed (i.e., when the union of all events in these processes is considered as a point process) the probabilistic law of the superposed process is, in general, unknown. For example, if the superposition of two independent renewal processes is a renewal process then all three processes are HPP's, Cinlar (1975); if the two processes being superposed are not HPP's then, not only is the superposed process not an HPP, in general it is not known what probabilistic law it obeys.

In many situations data are available for only a few systems since, e.g., in a reliability growth program during system development, only a few systems exist. In this situation, the PWP (1981) model cannot be used, nor can other techniques based on the distribution of time to first failure, the distribution of time between first and second failure, etc., be used. In addition, there may not be enough failures of individual systems to apply point process techniques to them. Under such circumstances, if analysis is to be performed at all—and certainly, it should be performed if at all possible—data must be pooled.

The problem of investigating the superposition of a small number of point processes, each observed over a relatively short time, is not readily subject to exact analysis. Quite likely, therefore, it will have to be approached by means of Monte Carlo simulation.

The results obtained by Jewell (1978a) serve to emphasize the need for investigations in this area. Jewell showed that plausible estimators of the Mean Time Between Failures *tended to increase* with increasing operating time when a sample path of a single *renewal* process was observed. This result remained true even for the special case of an HPP! Moreover, Jewell showed that pooling data for two or more systems made the situation even worse, i.e., the corresponding estimators for pooled data increased towards the asymptotic, true values *even more slowly.* Once again, this held true even under the HPP assumption! Most work on superposition of point processes has concentrated on the asymptotic case where either operating time or the number of pooled processes, or both, tend to infinity. Jewell's results highlight the need for investigations appropriate to reliability problems where very often both time and number of tested systems are small.

c. Conditions under which Superposed Renewal Processes Approach an HPP for Finite $t$

In subsection b above, we emphasized the need for examining the superposition of a small number of point processes, each observed for a short interval of time. Only a small number of processes were superposed since only a small number of systems were observed. When we investigate the probabilistic law governing the failure pattern of a single system, however, we can usually assume that a large number of renewal processes are being superposed since most systems consist of many parts. There have been a number of investigations of superposed renewal processes as the number of parts, $n$, approaches infinity. It has been shown, e.g., by Drenick (1960), that when $t = \infty$ and $n \to \infty$ the superposed process approaches an HPP. This result is of great interest since it provides a greatly simplified and unified result, analogous to the Central Limit Theorem (CLT) for sums of independent random variables. As with the CLT though, the assumptions which lead to the HPP result are not necessarily realistic. The assumption that $n$ is very large will usually be plausible but, rather than the asynchronous sampling implied by the $t = \infty$ assumption, most sockets of many systems might be considered to be sampled synchronously for the entire operational life of the system. That is, since many systems are retired from service with a large percentage of their original parts still installed, they never achieve the state where the effects of the boundary condition, that new parts are installed at $t = 0$, become negligible.

In light of the above discussion, a result due to Grigelionis (1964) should have received greater notice than it has. Put briefly, Grigelionis showed that under the following conditions—some of which are very realistic—the superposed process converges to an HPP even for small $t$. Consider a system of $n$ parts in series where each failed part is replaced instantaneously. In the $r^{\text{th}}$ component position, the initial component has life distribution $\hat{F}_{nr}$ while all subsequent components have life distribution $F_{nr}$; $r = 1, \ldots, n$. Let $X_{nr}(t)$ be the number of failures occuring in the $r^{\text{th}}$ part position during $(0,t]$ and $X_n(t) = \sum_{r=1}^{n} X_{nr}(t)$ be the total number of system failures during $(0,t]$. If it is assumed for fixed $t > 0$ that

$$\lim_{n\to\infty} \max_{1\leqslant r\leqslant n} \hat{F}_{nr}(t) = 0 = \lim_{n\to\infty} \max_{1\leqslant r\leqslant n} F_{nr}(t)$$

and

$$\lim_{n\to\infty} \sum_{r=1}^{n} \hat{F}_{nr}(t) = V(t)$$

then the counting process $\{X_n(t), 0 < t < \infty\}$ converges in probability to an NHPP whose expected number of failures in (0,t] is $V(t)$. Moreover, if $\hat{F}_{nr}(t)$ satisfies the requirement that $0 < \hat{F}_{nr}(0) < \infty$ then convergence will be to an HPP. These results of Grigelionis are discussed by Barlow and Proschan (1975, pp. 245-253), who state that, "It furnishes another important justification for the use of the exponential distribution." It is surprising that, to the best of our knowledge, there have not been any other references to Grigelionis's demonstration that the reliability model for a complex system, under appropriate assumptions, will be an HPP for all $t > 0$.

There are some practical problems, however, with the assumptions. We first want to stress, however, that the requirement that $0 < \hat{F}_{nr}(0) < \infty$ which ensures that convergence is to an HPP, is quite reasonable. This requirement is merely that the probability density function at time 0 is neither zero nor infinite. The former will often be especially realistic since for any cumulative distribution function, $F'(0) = 0$ implies that immediate failure is an impossible event. We note, however, that the distributions commonly used in reliability, i.e., the Weibull, gamma and log-normal distributions do not have this property. Since it is often physically plausible that $0 < F'(0) < \infty$, this is more of a reflection on these models than a valid criticism of the applicability of Grigelionis's theorem to the real world.

There are some problems, however, with other assumptions needed for convergence even to an NHPP. For example, the requirement that

$$\lim_{n\to\infty} \max_{1\leqslant r\leqslant n} \hat{F}_{nr}(t) = 0$$

for any fixed $t > 0$ means that, in effect, the least reliable part will never fail in any fixed time, even say, the total time from the "Big Bang" to the present. Of course, this requirement is tempered by the fact that

$$\lim_{n\to\infty} \sum_{r=1}^{n} \hat{F}_{nr}(t) = V(t)$$

and $V(t)$ can be arbitrarily large for large $t$. A more restrictive requirement is that (Barlow and Proschan (1975, p. 247, Eq. 4.5)) the following condition must hold

$$\lim_{n\to\infty} \sum_{r=1}^{n} Pr\{X_{nr}(t) \geqslant 2\} = 0$$

for fixed $t \geqslant 0$. To us it seems obvious that for real systems composed of real parts this sum will approach $n$ for large enough values of $t$ and $n$, of course, must be very large. (The requirement would be better met in a software reliability context where a properly installed fix will eliminate a program bug forever.) For small enough $t$ the above requirement will be adequately met but the superposition of renewal processes model itself is questionable for small $t$. Initially defective parts will be—at least in some cases—outliers from the population of parts from which a selection is made for the first installation. Moreover, design fixes are frequently installed early in the life of a system. Such fixes change system configuration, or as a minimum, the type(s) of part(s) installed in one or more sockets.

We recommend that a thorough investigation be made of the practical applicability of Grigelionis's results. For example, analyses along the lines of those performed by Blumenthal, Greenwood and Herbach (1971, 1973, 1976, 1978) should be undertaken. Blumenthal, et al. considered the superposition of finite numbers of renewal processes observed for finite times. However, since they dealt with the commonly used distributions, they did not treat the case where $0 < F'(0) < \infty$.

d. Competing Risks Problems

In many cases, there are two or more types or causes of failures to which either nonrepairable or repairable items are susceptible. The problem of greatest interest is to estimate the improvement in reliability if one or more of these failure causes are eliminated. In the case of nonrepairable items, the difficulty in estimating such improvement is notorious, see e.g., Kalbfleisch and Prentice (1980, pp. 177-178). The basic problem is that when only one failure can be observed on any one item, it is impossible to distinguish between independence and dependence of failure causes based only on data consisting of failure times and their causes. When multiple failures occur on a single item, the effect (or lack of effect) of one failure cause on one or more others can be observed. It is more straightforward, therefore, to estimate the effect of failure cause removal on the future reliability of a repairable system. Prentice, Williams and Peterson (1981) and Williams (1980, Chapter VI) consider this problem. Perhaps what is of greatest priority is implementation of their methods rather than further research.

e. Switchover from Ensemble Average to Point Process Approach

When we have no more than a few failures on each of a large number of systems it is clear that we would have to look across the ensemble of systems in order to perform meaningful analysis. Similarly, if we have a large number of failures on each of no more than a few systems, it is equally obvious that only the point process approach is feasible. Under what circumstances would we want to switch from one viewpoint to the other and over what set of conditions would we want to use both approaches to extract maximum information from the data? These questions are of great practical importance for the analysis of both reliability and biometry data but, to the best of our knowledge, they have never been addressed.

f. Alternatives to the NHPP for Reliability Growth Modeling

The NHPP has been used widely as a model for a system subject to deterioration, Barlow and Hunter (1960), Ascher (1968), on the basis that most repairs renew only a very small portion of a system. Therefore, the independent increments property of the NHPP (i.e., the numbers of failures in nonoverlapping intervals are independent random variables in an NHPP) will be satisfied, at least to an adequate approximation. The NHPP has also been applied to the modeling of reliability growth and in this situation the independent increments property is more questionable. In fact, the intent of a Test, Analyze and Fix (TAAF) Program is to test severly to find problem areas so that design fixes can be developed and installed to eliminate future occurrences of the same problems. In other words, the intent of a TAAF program is to keep increments from being independent. Moreover, it may be beneficial to model the process by which reliability growth occurs more closely than is possible with an NHPP. (On the other hand, it should be stated that the NHPP is a tractable model whose simplicity makes it appropriate for inclusion in Military Standards and Handbooks. Furthermore, as discussed in subsection b., pooling of data from two or more systems will often be required and there are open questions when this is done for models other than the NHPP.)

Braun and Schenker (1980), see also Braun and Paine (1977), have proposed the following alternative to the NHPP model:

$$\ln[v(t)] = d_0 + d_1 \ln [N_1(t) + 1] + d_2 u[N_2(t) + 1]$$

where

$$v(t) = \frac{d}{dt} \text{Expected Number of Failures} = \frac{d}{dt} [E \ N(t)]$$

$N(t) =$ Number of failures in $(0,t]$

$N_1(t) =$ Number of different failure modes discovered in $(0,t]$

$N_2(t) = N(t) - N_1(t)$, i. e., $N_2(t)$ is the number of failures in $(0,t]$ which are recurrences of previously observed modes.

$u(\cdot)$ is a function to be chosen. Braun and Schenker (1980) use three different functions: the identity, the square root and the natural logarithm

$d_0$, $d_1$ and $d_2$ are parameters.

The Braun and Schenker model requires three parameters, one more than in the commonly used power law process model

$$\rho(t) = \lambda \beta t^{\beta - 1} \quad .$$

Braun and Schenker fitted four models, their three (corresponding to their three choices for $u(\cdot)$ and the power law model to several sets of real and simulated data. In general, their model provided better forecasts, at the expense, of course, of an additional parameter. They recommended further work on their models on such topics as determining their stochastic properties and the development of standard errors of prediction.

It would be useful to investigate other models for reliability growth as well, including those originally introduced for software reliability. It has often been emphasized that a distinction between hardware and software reliability is that when a program bug is eliminated, that source of failure is removed forever. Perhaps too much has been made of this point, however, since in a reliable system composed of many parts, the reliability of each part is very high once outright design defects have been removed. Musa (1980) and Littlewood (1981) present reliability growth models which should be considered for hardware reliability.

### g. Probabilistic Modeling of Repairable Systems

Most modeling of repairable systems tends to be either overly simplistic or excessively complex and either approach may lead to unsatisfactory or unusable results. From the simplistic viewpoint, either a renewal process or an NHPP is often used as a systems model. While one or the other will sometimes be appropriate—for example, the NHPP is commonly accepted as a first order model for an automobile—these models do not necessarily provide enough detail or realism for satisfactory results. See our above comments on the use of the NHPP for reliability growth modeling (which, of course, is a special case of repairable system modeling).

At the other extreme, there have been a number of papers written which consider the intricacies of repairable systems with repairable subsystems, including the effects of finite repair times and a limited number of repairmen. Because of mathematical intractability, however, these papers usually assume that the studied systems are composed of only a very limited number of parts. Moreover, most results are stated under asymptotic conditions as operating time $\rightarrow \infty$, which is often very unrealistic. What is needed is an intermediate approach between these extremes which provide usable results for finite operating times, for complex systems which are neither renewed by each repair nor left completely aged after repair.

Cozzolino (1968) was a pioneering paper in the field of intermediate modeling of repairable systems. He focused on the situation where the rate of occurrence of failures was decreasing but his viewpoint can be generalized to deteriorating systems as well. It is of interest to note that Cozzolino's paper has seldom been referenced in the reliability literature and to the best of our knowledge, all such references are to his relatively brief treatment of nonrepairable items. The curious neglect of this paper highlights the general neglect of repairable systems.

Cozzolino considered three types of systems models, which he referred to as the "initial defects," the "N-component device" and the "time accumulation" models. We will not attempt to summarize these models here; instead we will point out a few of their realistic and unrealistic features.

In the "N-component device" model the age of each of a system's constituent parts is tracked making this model a very realistic representation of reality, at least in this regard. Additionally, the parts may have different distributions of time to failure under this model. A shortcoming of the model is its complexity, e.g., it has twice as many parameters as the number of system parts. In an attempt to reduce this complexity, Cozzolino introduced the time accumulation model where all the parts were identical so that one did not have to keep track of which part has failed.

In the time accumulation model, Cozzolino made the very reasonable assumption that replacement of one part in an $n$ part system resulted in the loss of $1/n$th of the system's age. This clearly is more realistic that the renewal assumption which requires complete loss of system age after replacement of $1/n$th of the system. It is also more realistic than the NHPP assumption that system age has not been reduced at all; for the first system failure the NHPP model is not too unrealistic but it becomes poorer as additional failures occur.

Cozzolino considered two versions of the "initial defects" model, one in which all defects had the same distribution of time to failure and another where there was variability in the failure properties of the defects. The specific assumptions on which this model is based include:

(1) Each new system has an unknown number, $M$, of initial defects.
(2) Each defect independently has an exponential density of time to failure. Under the homogeneous model, each of these exponential densities has the same parameter, $\lambda$, whereas under the other model, the parameters for the defects are independent samples from the same gamma distribution.
(3) The a priori distribution of the number of defects, $M$, is Poisson distributed.
(4) Each defect in the system will cause a failure, although not necessarily an immediate failure. Conversely, system failure can only occur because of a defect, i.e., failures do not occur because of any other causes.
(5) A defect, after causing a failure, is repaired perfectly, i.e., no additional failures can occur due to the same cause.

Note that, because the number of defects is Poisson distributed, there is a positive probability that $M = 0$, i.e., that the system has perfect reliability; there is also positive probability that $M$ exceeds any fixed number no matter how large. Under the above assumptions, Cozzolino showed that the system model was one of two NHPP's depending on the homogeneity or variability of the defects. One questionable result was that the rate of occurrence of failures decreased more rapidly under homogeneity than variability. One would expect that under variability, the defects with larger forces of mortality (hazard functions) would tend to occur sooner than the others, thus resulting in a relatively rapid decrease in the rate of occurrence of failures.

We strongly recommend that additional intermediate level modeling be performed. Cozzolino (1968), see also Cozzolino (1966), provides an excellent point of departure for such modeling. Of particular interest is Cozzolino's initial defects model which turns out to be the tractable NHPP model. How reasonable are the assumptions on which the initial defects model is based? What is the relationship between this model and the superposition models, discussed in subsection c.?

h. Locating Breakpoints on Repairable System Bathtub Curve

In the reliability field it is common practice to discuss *the* bathtub curve. In actuality, however, as first stressed by Krohn (1969), there are *two* bathtub curves. The shape of the plot is the same, but the ordinate for the bathtub curve for nonrepairable items is the force of mortality of a cumulative distribution function, whereas the ordinate for the curve for a repairable item is the rate of occurrence of failures of a stochastic point process. Ascher and Feingold (1981) discuss these bathtub *curves* in considerable detail; we stress, for example, that one of the chief reasons for this mixup is that in both cases the ordinate is usually called "failure rate."

Since bathtub models have been proposed for both nonrepairable and repairable items, it is useful to have statistical techniques to determine whether bathtub shaped forces of mortality or rates of occurrences of failures, respectively, are needed for adequate representations of reality. Given the existence of bathtub shaped models, it is also useful to estimate where the breakpoints on the curve occur. In the case of the force of mortality, appropriate techniques have been discussed by Shooman and Tenenbaum (1974) and Harris, Marchal and Zacks, (1973). Lee (1980) introduced a bathtub shaped rate of occurrence of failures and a recent paper by Farden (1980) may also be applicable to the problem of determining changepoints for NHPP's and, possibly, more complex models. Hinkley (1972) considered the situation where a sequence of IID RV's became a sequence of different IID RV's at a changepoint; his work may have some bearing on this problem.

It is often of great importance to be able to determine the time where system improvement has ceased and/or the time at which system deterioration commences. Even more basically, it is important to determine whether such changepoints exist in any particular situation. Further research in this area is certainly warranted.

i. Criteria for Adequacy of Representation of Repairable Systems Models

In the case of NHPP models, standard techniques, such as the Cramér-Von Mises and Kolmogorov-Smirnov goodness of fit (GOF) tests are applicable, see, e.g., Parzen (1962). It must be recognized, of course, that just as for distribution functions, these standard GOF techniques must be modified when parameters are estimated from data. Darling (1955) considered such modifications for the Cramér-Von Mises test and Crow (1974) applied Darling's techniques to the power law process

$$\rho(t) = \lambda \beta t^{\beta-1} .$$

Darling (1955) showed that his test, even in its limiting form, was not generally distribution free. That is, in general the limiting distribution (as sample size $\rightarrow \infty$ ) of his statistic usually depends on the structure of the family of distributions, $F(x;\theta)$, whose GOF to the data is being checked. Moreover, the limiting distribution, in general, depends on the true unknown value of the parameter $\theta$. In special cases, detailed by Darling (1955), the limiting distribution is independent of the value of $\theta$, which makes the test usable for large sample sizes.

There are still several outstanding questions. First, for classes of NHPP's other than the power law process, when will the limiting distribution be parameter free? Secondly, given this limiting property, what sample sizes are needed for the statistic to be parameter free, to an adequate approximation? (In practice, the required minimum sample size may be a function of the parameter value.) Thirdly, having established the criteria for obtaining usable results, these results will have to be calculated. Monte Carlo simulation will probably be required to obtain critical values of the test statistics.

The question of GOF criteria for point processes other than the HPP and NHPP is an open one. (Of course, in the case of a process known to be a renewal process, GOF techniques can be applied to the interarrival times of the process.) Some ad hoc procedures have been applied to point process and differential equation models, Schafer, Sallee and Torrez (1975), Braun and Paine (1977), Braun and Schenker (1980), but the results have not been totally satisfactory. In Ascher and Feingold (1981, Chapter 6) we indicated that two ad hoc GOF procedures were not capable of disclosing that data from deteriorating systems were being fitted with models that erroneously indicated growth. Moreover, these ad hoc procedures tracked poorly with the Darling GOF test when all three were applied to the power law process. In some cases, one or both of the ad hoc procedures indicated good fit when Darling's test strongly rejected the power law model. In other cases, the situation was reversed.

The existing GOF techniques have severe shortcomings, so better methods for comparing reliability growth models are needed. Note too, that if the performance of NHPP models is to be compared with other models, ad hoc procedures will have to be applied to the NHPP models as well. It is essential, therefore, to obtain improved GOF criteria for reliability growth models. Since some of these models are adapted to portraying deterioration, the flexibility of the models—and of GOF criteria developed for them—is enhanced.

## MILITARY STANDARDS AND HANDBOOKS

In the previous section we recommended research topics which will be useful in providing improved modeling and analysis of repairable systems. Here we will point out the need for improved documentation to help acclimate the reliability community to existing techniques for such systems.

### a. Military Handbook on Repairable Systems

In our previous papers we have emphasized the need for improved understanding of repairable system concepts, models and especially data analysis techniques. Much of this material is available to reliability personnel—at least in principle. Cox and Lewis (1966, p. 7) for example, emphasize that, "Industrial failure data form one of the most important fields of application for the methods of this monograph," but their text is almost unknown in the reliability field. Important papers, e.g., Jewell's (1978a,b) work showing that short term observation of a renewal process—and even of an HPP—will tend to indicate reliability growth and Prentice, Williams and Petersons's (1981) technique which is applicable to the common situation where many systems are under observation but most, or all, have suffered only a few failures, should also be brought to the attention of reliability personnel. All this material should be assembled under one cover and the word "Reliability" should be in the title of this document; we are convinced that the lack of this word in Cox and Lewis's (1966) title ("The Statistical Analysis of Series of Events") has been a major cause of this text's lack of influence in the reliability field.

In addition to the need to present repairable system techniques, it is crucial to introduce consistent and intuitively appropriate terminology and notation. The lack of appropriate terminology has been a principal contributor to the situation where, for example, the first formal statistical test ever developed, see Bartholomew (1955), is not discussed in a single reliability text. This is in spite of the fact that this test is very simply implemented and interpreted and is optimal for testing an HPP against reliability growth/deterioration, as modeled by at least two plausible alternative hypotheses!

A Military Handbook will provide the means for presenting techniques and terminology for repairable systems in a highly visible, formal manner. We strongly recommend the preparation of such a Handbook. One possible basis for the Handbook would be our monograph, Ascher and Feingold (1981).

b. Updating of MIL-STD-757

MIL-STD-757 supplies techniques for the evaluation of repairable system failure data. The a priori assumption is made, however, that the times between successive failures of a system are IID exponential. This Standard should be revised to include techniques to test for nonstationarity of successive times between failures of a system—where reliability growth is sought it is hoped that these successive times are *not* identically distributed. Techniques should be provided for the fitting of NHPP models to data shown to be nonstationary. Thought should also be given to include techniques for testing for independence of successive times between failures. It should be recognized, however, that the large number of failures of a single system required to perform tests for independence make it questionable whether such tests should be included. Moreover, the selection and fitting of models which account for dependence is quite intricate and would be difficult to formalize in a Military Standard.

c. Updating of MIL-STD-721

In our previous papers we have emphasized the need for improved, more consistent and intuitively appealing terminology and notation. We attempted to promote such improvements during the Electronic System Reliability Workshop in 1975 by recommending the updating of MIL-STD-721B, the standard for reliability definitions. In spite of our efforts during the Workshop and during the development of MIL-STD-721C, this latest version of the Standard does not include *any* terms for repairable systems. As indicated in the vicious cycle in the Introduction to this paper, there are such widespread and deep-seated confusions in concepts—largely due to poor terminology—that it is extremely difficult to demonstrate the need for better terminology. If some headway is made in clarifying this area by other means— e.g., through the introduction of the Military Handbook on repairable systems discussed above—it will become feasible to incorporate repairable systems definitions in a future version of MIL-STD-721.

## ACKNOWLEDGMENT

The support of the Office of Naval Research, particularly Mr. Seymour Selig, is gratefully acknowledged.

## REFERENCES

H.E. Ascher (1968), "Evaluation of Repairable System Reliability Using the 'Bad-As-Old' Concept," IEEE Trans., R-17, 103-110.

H.E. Ascher and H. Feingold (1969), "'Bad-As-Old' Analysis of System Failure Data," in Annals of Assurance Sciences, Gordon and Breach, New York, pp. 49-62.

H.E. Ascher and H. Feingold (1978a), "Is There Repair after Failure?," in ARMS*, IEEE-77CH1308-6R, pp. 190-197.

H.E. Ascher and H. Feingold (1978b), "Application of Laplace's Test to Repairable System Reliability," in Proc. Int. Conf. on Reliability and Maintainability, Société Pour La Diffusion Des Sciences Et Des Arts, France, pp. 219-225.

H.E. Ascher and H. Feingold (1979), "The Aircraft Air Conditioner Data Revisited," in ARMS*, IEEE-79CH1429-OR, pp. 153-159.

H.E. Ascher and H. Feingold (1981), "Repairable Systems Reliability: Modeling, Inference, Misconceptions and Their Causes," to be published by Marcel Dekker, New York.

R.E. Barlow and L. Hunter (1960), "Optimum Preventive Maintenance Policies," Operations Res., 8, 90-100.

R.E. Barlow and F. Proschan (1975), "Statistical Theory of Reliability and Life Testing Probability Models," Holt, Rinehart and Winston, New York.

---

*ARMS ≡ Annual Reliability and Maintainability Symposium

D.J. Bartholomew (1955), Discussion of D.R. Cox, "Some Statistical Methods Connected with Series of Events," J. Roy. Stat. Soc., Ser. B., 17, pp. 162-163.

S.B. Blumenthal, J.A. Greenwood and L.H. Herbach (1971) "Superimposed Non-stationary Renewal Processes," J. of Appl. Prob., 8, 184-192.

S.B. Blumenthal, J. A. Greenwood and L.H. Herbach (1973), "The Transient Reliability Behavior of Series Systems or Superimposed Renewal Processes," Technometrics, 15, 255-269.

S.B. Blumenthal, J.A. Greenwood and L.H. Herbach (1976), "A Comparison of the Bad-As-Old and Superimposed Renewal Models," Management Science, 23, 280-285.

S.B. Blumenthal, J.A. Greenwood and L.H. Herbach (1978), "The Curse of the Exponential Distribution in Reliability," Proc. Twenty-Third Conf. on the Design of Experiments in Army Research Development and Testing, ARO Report 78-2, pp. 457-471.

H. Braun and J.M. Paine (1977), "A Comparative Study of Models for Reliability Growth," Dept. of Stat., Princeton Univ., Tech. Rep. No. 126, Series 2.

H. Braun and N. Schenker (1980), "New Models for Reliability Growth," Dept. of Stat., Princeton Univ., Tech. Rep. No. 174.

E. Çinlar (1975), "Introduction to Stochastic Processes," Prentice-Hall, Englewood Cliffs, N.J.

D.R. Cox (1972), "Regression Models and Life Tables (with Discussion)," J. Roy. Stat. Soc., Ser. B., 34, 187-220.

D.R. Cox and P.A. Lewis (1966), "The Statistical Analysis of Series of Events," Methuen, London.

J.M. Cozzolino, Jr. (1966), "The Optimal Burn-In Testing of Repairable Equipment," Operations Research Center, MIT, Tech. Rep. No. 23.

J.M. Cozzolino, Jr. (1968), "Probabilistic Models of Decreasing Failure Rate Processes," Nav. Res. Logistics Quarterly, 15, 361-374.

L.H. Crow (1974), "Reliability Analysis for Complex Repairable Systems," in "Reliability and Biometry," F. Proschan and R.J. Serfling, eds., SIAM, Philadelphia, pp. 379-410.

D.A. Darling (1955), "The Cramér-Smirnov Test in the Parametric Case," Ann. Math. Stat., 26, 1-20.

R.F. Drenick (1960), "The Failure Law of Complex Equipment," J. Soc. Indust. Appl. Math., 8, 680-690.

N.J. Farden (1980), "Testing for a Change Point," Colorado State Univ., Stat. Dept. Rep. No. 19.

B.I. Grigelionis (1964), "Limit Theorems for Sums of Renewal Processes," in "Cybernetics in the Service of Communism," Vol. 2, Reliability Theory and Queueing Theory, A.I. Berg, N.G. Bruevich and B.V. Gnedenko, eds., "Energy" Publishing House, Moscow-Leningrad, pp. 246-266.

C.M. Harris, W.G. Marchal and S. Zacks (1973), "Failure-Rate Prediction and Wearout Detection," School of Eng. and Appl. Sci., George Washington Univ., Serial T-282.

D.V. Hinkley (1972), "Time-ordered Classification," Biometrika, 59, 509-523.

W.S. Jewell (1978a), "'Reliability Growth' as an Artifact of Renewal Testing," Operations Research Center, Univ. of California, Berkeley, ORC 78-9.

W.S. Jewell (1978b), "A Curious Renewal Process Average," Operations Research Center, Univ. of California, Berkeley, ORC 78-12.

J.D. Kalbfleisch and R.L. Prentice (1980), "The Statistical Analysis of Failure Time Data," John Wiley, New York.

C.A. Krohn (1969), "Hazard Versus Renewal Rate of Electronic Items," IEEE Trans., R-18, 64-73.

L. Lee (1980), "Testing Adequacy of the Weibull and Log Linear Rate Models for a Poisson Process," Technometrics, 22, 195-199.

B. Littlewood (1981), "Stochastic Reliability Growth: A Model for Fault-Removal in Computer-Programs and Hardware Designs," IEEE Trans., R-30, 313-320.

J.D. Musa (1980), "The Measurement and Management of Software Reliability," Proceedings of IEEE, 68, 1131-1143.

E. Parzen (1962), "Stochastic Processes," Holden-Day, San Francisco.

R.L. Prentice, B.J. Williams and A.V. Peterson (1981), "On the Regression Analysis of Multivariate Failure Time Data," Biometrika, 68, 373-379.

R.E. Schafer, R.B. Sallee and J.D. Torrez (1975), "Reliability Growth Study," Hughes Aircraft Co., RADC-TR-75-253.

M.L. Shooman and S. Tenenbaum (1974), "Hazard Function Monitoring of Airline Components," in ARMS*, IEEE 74CH0820-1RQC, pp. 383-390.

W.A. Thompson, Jr. (1981), "On the Foundations of Reliability," Technometrics, 23, 1-13.

B.J. Williams (1980), "Proportional Intensity Models for Multiple Event Times," Unpublished Ph.D. Dissertation, University of Washington.

# Statistical Estimation, Using Real Data from Systems Having a Decreasing Hazard Rate, and its Application to Reliability Improvement

Sam C. Saunders

*Washington State University*

**Abstract**

In a recent SECNAV Notice 5000, signed by D. E. Mann, on the subject *Major Systems Acquisitions* is the following paragraph: "Although there is considerable uncertainty early in the acquisition process, *every effort must be made to use the best available data and techniques in developing estimates* (emphasis added). Bands of uncertainty shall be identified for point estimates. Broad bands of uncertainty shall be expected early in the acquisition process, with smaller bands developed as the program matures and the uncertainty decreases."

For the class of failures which are exponential, i.e., have constant hazard (failure) rate, the theory and methods for estimation are now well known (witness the use of such terms as MTBF, mean time between failures, loc. cit.) and are correct when life has a constant hazard rate.

What is undertaken here is an exposition of a theory and method of parametric estimation in the increasingly important situation namely when the hazard rate decreases with age; not that each component necessarily improves with age but each component in a group may have a different, but constant, hazard rate in service and so for older components it becomes more likely that its hazard rate is small.

## Technical Summary

Hazard rates, which decrease from an initial value of $\theta + \frac{\lambda}{\phi}$ to an ultimate

Research, in part, supported by Office of Naval Research N00014-79-C-0755 value of $\theta$, of the form

$$h(t) = \theta + \lambda/(\phi + t) \text{ for } t > 0,$$

are shown to arise in practical engineering situations. Such failure rate and their estimation were first discussed by Davis and Feldstein (1979) in a biological application.

After reparameterization, maximum likelihood estimates are obtained as the implicit solution of certain equations using results of Saunders and Myhre (1980) even in the case there is a paucity of failure observations among those observed. Computational algorithms are devised which can be used on a programmable calculator such as HP-41C.

The utility of these estimation procedures is shown by confronting the theory with several sets of actual system life data as a test of adequacy. While the actual numerical estimates are not computed herein, the computational methods to obtain the estimators are discussed. The advantage of having such parametric estimates available is demonstrated by showing how they can be used in the computation of optimal burn-in periods (preconditioning) for such systems.

A method to obtain approximation confidence intervals in the case $\theta = 0$, is adopted from Saunders and Myhre (1981) and shown to apply to the reliability of such systems even when the data does not have a large number of failures.

## 1. INTRODUCTION

Let us consider a system which has a decreasing hazard rate $h$ defined by

$$h(t) = \frac{\alpha\beta}{1 + t\beta} + \alpha\beta\gamma \text{ for } t > 0 \tag{1.1}$$

where the parameters $\alpha, \beta, \gamma > 0$ are unknown. See Figure 1.

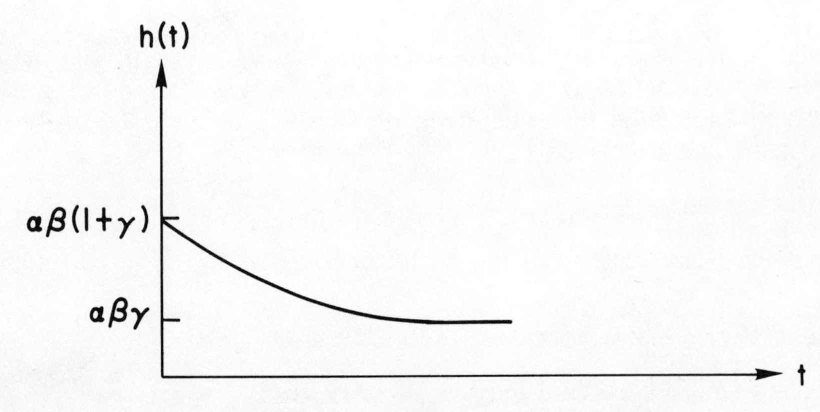


Figure 1

The hazard function $H$ is given by

$$H(t) = \alpha[\ln(1 + t\beta) + \beta\gamma t] \text{ for } t > 0. \tag{1.2}$$

The hazard rate here is the sum of two terms: the first is a convex decreasing term and the second a constant term. It is a reparameterization of a generalized Pareto law studies by David and Feldstein (1979) and given by them in the form

$$h(t) = \theta + \lambda/(t + \phi). \tag{1.3}$$

By setting $\theta = \alpha\beta\gamma$, $\lambda = \alpha$, $\phi = 1/\beta$ we obtain (1.1). In the form given in (1.3) only the restrictions $\lambda \geqslant -\theta\phi$, $\theta$, $\phi > 0$ need be observed, so that (concave) increasing hazard rates are possible. Such is not the case with (1.1) and the restrictions given there.

While concave increasing hazard rates may be of interest in the study of the etiology of a disease, as was the case in the paper cited, they are not of primary interest in engineering application where convex hazard rates are usually observed. For our subsequent use the parameterization given in (1.1) is assumed.

Davis and Feldstein (1979) gave an estimator of $\phi$ based on the Grenander (1956) maximum likelihood estimate of a monotone hazard rate and consequently it depends only upon the failure times which appear in the sample. Then presuming $\phi$ to be known they then provided equations implicitly defining the MLE's of $\theta$ and $\lambda$ under progressively censored sampling and computational methods for their evaluation.

Such a procedure might not be sufficiently reliable in the case when there is a paucity of failure observations in the sample, e.g., say two or three failures among dozens of observations. Unfortunately this is often the case in engineering studies, since the entire effort of design is to reduce the number of failures as close to zero as possible.

Saunders and Myhre (1981) have provided maximum likelihood estimators and computational procedures, under such realistic sampling conditions, for a class of decreasing hazard rate distributions with two unknown parameters of the form

$$H(t) = \alpha\ Q(t\beta) \tag{1.4}$$

when certain assumption are made about the behavior of $q = Q'$, namely for the functions $\zeta$ and $\psi$ defined by

$$\psi(x) = xq(x),\ \zeta(x) = 1 + xq'(x)/q(x) \tag{1.5}$$

satisfy

1° $\psi$ is increasing
2° $q$ is log-convex
3° $\zeta$ is bounded between 0 and $1 - \zeta$ is unimodal.

## 2. THE GENESIS OF THE MODEL

Let us assume that each system, when exposed to evaluation testing, will have a constant hazard (or failure) rate, say $\lambda$, and thus its life until failure will be an exponential random variable. But due to variation between vendors, or in manufacturing quality control, the failure rates of all such systems will vary slightly in some random manner; presume this variation has a $\Gamma(\alpha,\beta)$ distribution.

Then the reliability of any such system, selected at random to be put under test evaluation, will be given by

$$R(t) = \int_0^\infty e^{-\lambda t} g(\lambda)\, d\lambda = \exp\{-\alpha \ln(1 + t\beta)\} \tag{2.1}$$

where the gamma density is, for $\alpha > 0$, $\beta > 0$, defined by

$$g(\lambda) = \frac{\lambda^{\alpha-1}\, e^{-\lambda/\beta}}{\beta^\alpha\, \Gamma(\alpha)} \quad \text{for } \lambda > 0.$$

The failure rate of the system then will be

$$q(t) = \frac{\alpha\beta}{1 + t\beta} \quad \text{for } t > 0. \tag{2.2}$$

Here the choice of the gamma as the mixing distribution comes from two considerations. Firstly, the gamma distribution is a flexible two-parameter family capable of representing many situations adequately and secondly, it yields a closed-form solution, as in (2.1), which has been encountered previously and is called the Lomax or Pareto type II distribution.

The hazard rate given in (2.2) is decreasing, as all such mixed-exponential distributions will be. Samples from such a distribution will often be highly censored because of the expense of testing and since, if a system does not fail early, it is judged "good" and probably will not fail during the test. Samples from two flight control electronic packages are given in data sets I and II.

A distribution with hazard rate of the form (1.3) can arise when a system having reliability given (2.1) whenever it is subjected to laboratory test and evaluation, is put into a field test or actual service and then subjected to possible catastrophic accidents which have a constant rate of occurrence independent of the failure distribution generated by manufacturing variability as represented in (2.1).

Consequently the life of such a system in service will have a survival distribution which is the product of the two survival probabilities, namely

$$\bar{F}(t) = R(t)\, e^{-\theta t} = \exp\{-\alpha \ln(1 + t\beta) - \theta t\} \tag{2.3}$$

Thus by reparameterization of (2.3) we obtain the hazard rate as given in (1.2).

This can be regarded as a failure model in which both infant mortality and the constant mortality rate of mid-life accidents are present but there is no wear-out i.e., no terminally increasing hazard rates. The assumptions of this mathematical explanation are thought to be particularly appropriate for electronic equipment containing subsystems which are integrated circuits.

## 3. MAXIMUM LIKELIHOOD ESTIMATION FOR PROGRESSIVELY CENSORED DATA

Denote the sample data with the vector $\underline{t} = (t_1, \ldots, t_k, t_{k+1}, \ldots, t_n)$ where $t_1, \ldots, t_k$ are the ordered failure observations and $(t_{k+1}, \ldots, t_n)$ are the ordered censoring times. In this case the log-likelihood from (1.1) and (1.2) is merely

$$L(\alpha, \beta, \gamma | \underline{t}) = \frac{1}{k} \sum_{1}^{k} \ln h(t_i) - \frac{1}{k} \sum_{1}^{n} H(t_j) \tag{3.1}$$

which we alter for our purpose by defining

$$H(t) = \alpha Q(t\beta{:}\gamma) \quad h(t) = \alpha\beta q(t\beta{:}\gamma)$$

where

$$Q(t{:}\gamma) = \ln(1 + t) + \gamma t, \quad q(t{:}\gamma) = \gamma + 1/(1 + t).$$

Letting $F_k$ denote the empirical cumulative distribution associated with $(t_1, \ldots, t_k)$ and $F_n$ denote that associated with $\underline{t}$ and $W_n$ the distribution with density defined by

$$W'_n(t) = [1 - F_n(t)]/\bar{t} \text{ for } t > 0 \tag{3.2}$$

then for any function $g$ we define the corresponding transforms

$$\bar{g}(x) = \int_0^\infty g(tx)\, dF_n(t), \quad \tilde{g}(x) = \int_0^\infty g(tx)\, dF_k(x)$$

and similarly $g^{\#}(x) = \int_0^\infty g(tx)\, dW_n(t)$.

Thus we have, utilizing the same notation as in (1.5), the stationary points of the likelihood given by the simultaneous solution to the equations:

$$\frac{\partial L}{\partial \alpha} = 0 \text{ iff } \frac{1}{\alpha} = \frac{n}{k} \bar{Q}\,(\beta{:}\gamma), \tag{3.3}$$

$$\frac{1}{\beta}\frac{\partial L}{\partial \beta} = 0 \text{ iff } \tilde{\zeta}(\beta{:}\gamma) = \frac{\alpha n}{k} \bar{\psi}\,(\beta{:}\gamma), \tag{3.4}$$

$$\frac{\partial L}{\partial \gamma} = 0 \text{ iff } \frac{1}{k} \sum_{i=1}^{k} [(1 + t_i\beta)^{-1} + \gamma]^{-1} = \frac{\alpha\beta n}{k}\, \bar{t}. \tag{3.5}$$

Now we will solve these three equations iteratively in the following fashion for $i = 1, 2, \ldots$

(1) Given $\alpha_{i-1}, \beta_{i-1}$ substitute them into (3.5). Then solve as a function of $\gamma$, call nonnegative root $\gamma_i$. (This can be done easily using Newton's method).

(2) Solve the two Eqs. (3.3) and (3.4) for $\alpha, \beta$ with $\gamma$ fixed at $\gamma_i$ in the same manner as described, loc. cit., namely find $\beta_i$ the smallest positive solution of the equation

$$\tilde{\zeta}(\beta:\gamma_i) = \frac{\bar{\psi}(\beta:\gamma_i)}{\bar{Q}(\beta:\gamma_i)}. \tag{3.6}$$

Having done so we then compute

$$\alpha_i = k/n\ \bar{Q}(\beta_i:\gamma_i). \tag{3.7}$$

To assure ourselves that these equations can be solved in the manner indicated we first examine Eq. (3.5) which can be written $f(x) = c$ for some constant $c$ where for given $\beta > 0$

$$f(x) = \frac{1}{k} \sum_{i=1}^{k} [(1 + t_i\beta)^{-1} + x]^{-1}.$$

Note that $f'(x) < 0$ and $f''(x) > 0$, so $f$ is convex decreasing and $f(\infty) = 0$. Thus a unique positive solution always exists iff $f(0) > c$. Otherwise we take the solution $x = 0$. We call the root $\gamma$.

We consider the remaining two Eqs. (3.3) and (3.4) with $\gamma$ fixed. We suppress this dependence upon $\gamma$ in our notation. Eliminating $\alpha$ from the two equations, we must solve Eq. (3.6) in the variable $\beta$.

To determine the existence and difficulty of finding the least positive root, if more than one exists, we examine the likelihood from (3.1) with $\gamma$ fixed:

$$L(\alpha,\beta|\gamma,\underline{t}) = \ln(\alpha\beta) + \widetilde{\ln q}(\beta) - \frac{\alpha n}{k}\bar{Q}(\beta).$$

Maximizing this with respect to $\alpha$ yields a modified likelihood function of $\beta$:

$$\begin{aligned} L(\beta) = L[(k)/n\bar{Q}(\beta),\ \beta|\gamma,\ \underline{t}] &= \ln\left(\frac{k}{n}\right) - 1 + \widetilde{\ln q}(\beta) - \ln\left(\frac{\bar{Q}(\beta)}{\beta}\right) \\ &= \widetilde{\ln q}(\beta) - \ln[q^{\#}(\beta)] + \text{constants independent of } \beta. \end{aligned}$$

Since, as before integrating by parts, we find

$$\frac{\bar{Q}(\beta)}{\beta} = \bar{t}\int_0^\infty q(t\beta)\ dW_n(t) \equiv \bar{t}\cdot q^{\#}(\beta)$$

where $W_n$ is the distribution with density given in (3.2).

One checks that $\beta L'(\beta) = 0$ is the same as (3.6) so that independent of $\gamma$, the smallest positive root is a maximum of $L$ iff $L$ is positive in a neighborhood of zero iff $\tilde{t} < t^{\#} = \overline{t^2}/2\bar{t}$. This inequality is a necessary condition that a MLE exists.

## 4. OPTIMUM BURN-IN AND ULTIMATE HAZARD RATE

For a system with hazard rate given by (1.1), a burn-in of length $\tau \geqslant 0$ will yield a (random) remaining life $X_\tau$ with a hazard rate in the same family but with different parameters. It is given by

$$h(t+\tau) = \frac{\alpha\beta}{1+(t+\tau)\beta} + \alpha\beta\gamma = \alpha\beta' q(t\beta' : \gamma')$$

where

$$\beta' = \beta/(1+\tau\beta), \ \gamma' = \gamma(1+\tau\beta).$$

After the burn-in the initial hazard rate is lowered to $\alpha\beta'(1+\tau')$ from the value $\alpha\beta(1+\gamma)$, while the terminal hazard rate remains $\alpha\beta'\tau' = \alpha\beta\gamma$. Given that three parameters have been estimated by the methods of section 3 for such a system, we ask what benefits can be calculated from this knowledge? That depends upon the criterion of benefit.

Criterion I: If the cost per unit time of burn-in is \$C and the benefit is \$B for each failure per hour of the maximum failure rate reduced by the burn-in, then the increased gain per system for a burn-in of duration $\tau$ is

$$\begin{aligned} g(\tau) &= B[\alpha\beta(1+\gamma) - \alpha\beta'(1+\gamma')] - C\tau \\ &= \alpha\beta B[1 - \frac{1}{1+\tau\beta}] - C\tau. \end{aligned}$$

To maximize our gain we solve for $\tau$ in $g'(\tau) = 0$ or

$$\alpha\beta B \frac{\beta}{(1+\tau\beta)^2} = C.$$

Since a burn-in can not be negative, we can see the optimum burn-in for the criterion I above is

$$\tau = \left(\frac{\sqrt{\alpha B}}{C} - \frac{1}{\beta}\right)^+ \quad \text{where } x^+ = \max(x, 0).$$

Criterion II: What will it cost to bring the initial hazard rate to within 100 p% of the ultimate hazard rate using a green-run? We want to have

$$\alpha\beta'(1+\gamma') = \alpha\beta\gamma(1+p).$$

Solving for $\tau$ we find

$$\tau = \frac{1}{\beta}\left(\frac{1}{\gamma p} - 1\right)^+.$$

Suppose instead that the total money available for a burn-in is \$D per system. How close, to within what percentage, can we bring the initial hazard rate after burn-in to the ultimate value which can be obtained?

The percentage of initial to terminal hazard rate for a burn-in a length $\tau$ is

$$\frac{\alpha\beta'\,(1+\gamma)}{\alpha\beta'\gamma'} = 1 + \frac{1}{\gamma(1+\tau\beta)}.$$

Since the number of units of time that the burn-in can utilize is $D/C = \tau$. Hence the percentage is

$$[\gamma(1 + \frac{D}{C}\beta)]^{-1}.$$

Criterion III: Suppose the benefit due to increased reliability is \$B per unit time of increased expected life at a cost of \$C per unit time of burn-in. Recall $X_\tau$ is the life after burn-in of length $\tau$. Then the gain as a function of $\tau$ becomes

$$g(\tau) = B\;E(X_\tau - X_0) - C \cdot \tau$$

with maximum determined from

$$g'(\tau) = B\;\partial\frac{EX_\tau}{\partial\tau} - C = 0.$$

Thus we must find $\tau$ such that

$$\frac{\partial EX_\tau}{\partial\tau} = C/B.$$

Now we recall the formula for mean

$$EX_\tau = \int_0^\infty \exp\,[-H(t+\tau) + H(\tau)]\,dt.$$

Hence

$$\begin{aligned}\frac{\partial EX_\tau}{\partial\tau} &= \int_0^\infty [h(\tau) - h(t+\tau)]\,\exp\{H(\tau) - H(t+\tau)\}\,dt\\ &= h(\tau)e^{H(\tau)}\int_\tau^\infty e^{-H(t)}\,dt - 1.\end{aligned}$$

Thus we must, using criterion III, solve for $\tau$ in the equation

$$h(\tau)\;e^{H(\tau)}\int_\tau^\infty e^{-H(\tau)}\,dt = \frac{B+C}{B}.$$

Unfortunately this becomes a problem involving numerical integration and perhaps the use of Newton-Raphson iteration.

Of course, the actual cost of a burn-in may also include set-up costs in which case the criterion for optimality will become even more complicated numerically. But with the estimates of the parameter available from the preceding analysis, it becomes a problem which can be handled by machine calculation.

Some actual data sets from electronic package testing are now presented. These illustrate the size of the samples and the degree of censoring which must be accommodated by any practical theory.

Data Set I: Time in minutes

| | Failed | Censored | | | |
|---|---|---|---|---|---|
| $k = 3$ | 1 | 59 | 113 | 145 | 182 |
| $n = 14$ | 8 | 72 | 117 | 149 | 320 |
| | 10 | 76 | 124 | 153 | |

Data Set II: Time in minutes

| | Failed | Censored | | | |
|---|---|---|---|---|---|
| $k = 2$ | 37 | 60 | 66 | 72 | 123 |
| $n = 9$ | 53 | 64 | 70 | 96 | |

Data Set III: Time in hours Pat & Pre-Pomp

| | Failed | | Censored | | | |
|---|---|---|---|---|---|---|
| | .10, | .43 | 2.35 | 2.63 | 3.18 | 4.50 |
| $k = 10$ | .11, | 1.17 | 2.37 | 2.87 | 3.23 | 5.12 |
| $n = 42$ | .20, | 1.50 | 2.43 | 2.88 | 3.24 | 5.42 |
| | .27, | 1.98 | 2.48 | 2.92 | 3.45 | 6.82 |
| | .37, | 2.32 | 2.50 | 3.00 | 3.75 | 8.15 |
| | | | 2.52 | 3.07 | 3.87 | 11.15 |
| | | | 2.57 | 3.08 | 3.93 | 11.30 |
| | | | 2.58 | 3.10 | 3.97 | |
| | | | 2.63 | | | |

Data Set IV: Time in hours Init. SLE

| | Failed | Censored | | | | |
|---|---|---|---|---|---|---|
| | .18 | 1.48 | 2.90 | 4.30 | 4.55 | 5.18 |
| $k = 7$ | .50 | 1.48 | 3.53 | 4.33 | 4.60 | 5.30 |
| $n = 42$ | .70 | 2.18 | 3.63 | 4.46 | 4.95 | 5.45 |
| | .90 | 2.25 | 3.75 | 4.52 | 5.00 | 5.47 |
| | .95 | 2.63 | 4.30 | 4.55 | 5.00 | 5.85 |
| | 1.25 | | | | | |
| | 6.83 | | | | | |

## 5. LOWER CONFIDENCE BOUNDS ON THE RELIABILITY OF SUCH SYSTEMS

Part of the requirements for the acquisition of systems is that confidence bounds on their reliability must be supplied at each "milestone" review. In particular, lower confidence bounds are often of primary interest.

Let us assume that we have life length $X$, with survival distribution

$$\bar{F}_X(y) = e^{-\alpha Q(\beta y)} \text{ for } y > 0 \tag{5.1}$$

where the shape and scale parameters $\alpha, \beta$ are unknown and $Q$ is a known concave increasing hazard function with hazard rate satisfying 1°, 2°, and 3°.

However, it is supposed that the random variable that can be observed is

$$Y = \min(X, T)$$

where $T$ is a censoring time, the distribution of which is unknown, but it is independent of $X$. This means that an observation is either a failure or the test is terminated without failure for some reason which can be end-of-day, end-of-funds, etc. but which is regarded as an observed value of $T$.

In Saunders and Myhre (1981), it is shown under mild conditions, e.g. if $T$ has support only on a bounded region, that $(\hat{\alpha}, \hat{\beta})$, the MLE's from large randomly censored samples, are strongly consistent and asymptotic bivariate normal with mean $(\alpha, \beta)$ and symmetric covariance matrix $\frac{1}{n} \Sigma^{-1}$ where

$$\Sigma = \begin{bmatrix} \frac{p}{\alpha}, & \int_0^\infty yq(y\beta)\, dF_Y(y) \\ , & \frac{\alpha}{\beta} \int_0^\infty \zeta^2(y\beta)\, q(y\beta)\, \bar{F}_Y(y)\, dy \end{bmatrix}, \tag{5.2}$$

here $p = P[X < T]$.

In the case we have a sample vector $\underline{t} = (t_1, \ldots, t_k, \ldots, t_n)$ consisting of $k$ failures and $(n - k)$ censored values with $n$ large and each observation is assumed to come from distribution $F_Y$, then we can substitute the empiric cumulative $F_n$ for $F_Y$ in (5.2) to obtain an estimate of the matrix of the quadratic form of $(\hat{\alpha}, \hat{\beta})$.

$$\begin{bmatrix} \frac{k}{n\alpha}, & \frac{1}{\beta}\bar{\psi}(\beta) \\ , & \frac{\alpha \bar{t}}{\beta} (\zeta^2 \cdot q)^{\#}(\beta) \end{bmatrix}$$

where we have made use of the notation previously introduced in (3.2).

To supply a confidence interval for the estimated reliability

$$\hat{R}(y) = \exp\{-\hat{\alpha}\, Q(\hat{\beta} y)\} \text{ for } y > 0$$

we merely make use of a Taylor's series bivariate expansion of the hazard function to compute the variance of the asymptotically normal variate $\hat{\alpha} Q(\hat{\beta} y)$ with asymptotic mean $\alpha Q(\beta y)$ and then proceed to evaluate a lower bound at the approximate level of confidence specified. This procedure may be quite satisfactory in many instances, see Myhre and Saunders (1981).

The construction of exact confidence bounds in these instances is the subject of a separate study.

## BIBLIOGRAPHY

Davis, H.T. and Feldstein, M.L. (1979) The generalized Pareto Law as a model for progressively censored survival data, Biometrika, 66, 299-306.

Saunders, S.C. and Myhre, J.M. (1980) Maximum likelihood estimation for two parameter decreasing failure rate distributions using censored data, Washington State University, Math. Dept. Technical Report TR-80-3.

Myhre, J.M. and Saunders, S.C. (1981) On the asymptotic behavior of certain maximum likelihood estimators from large, randomly censored samples, Washington State University Math. Dept. Technical Report, TR-81-1.

# Part III: Reliability Life Testing

# Some Current Research in Reliability: An Overview

Nozer D. Singpurwalla

*The George Washington University*

**Abstract**

This paper is based upon a talk given by the author at the Office of Naval Research/Army Research Office Reliability Workshop held in Washington, D.C. on April 29 — May 1, 1981. The ONR portion of the workshop featured "Reliability in the Acquisitions Process" as a theme, and this paper is a state of the art overview of some research appropriate to this theme. The topics discussed here are labelled as follows: (i) Robustness of the MIL-STD-781C sequential life testing procedures, (ii) A model for step-stress testing, (iii) A binary response damage prediction problem, and (iv) The stochastic characterization of life lengths generated by a minimal repair policy.

The work described under all of the above topics has been motivated by genuinely practical considerations, and only those results which have direct practical use have been reported. Furthermore, with the interest of potential users in mind, detailed proofs and discussions that support the results have been omitted. These are available in the appropriate papers, which are referenced in the text of this paper.

## 1. ROBUSTNESS OF THE MIL-STD-781C SEQUENTIAL LIFE TESTING PROCEDURES

A sequential procedure to test hypotheses about the mean life of an exponential distribution is widely used in government and industry. This procedure is due to Epstein and Sobel (1955), and is codified in MIL-STD-781C. Often in practice, MIL-STD-781C is also used when the underlying distribution is *not* exponential. Our goal in this research is to investigate the consequences of this misuse, especially its effect on the producer's risk, the consumer's risk, and the expected sample size. Such an investigation is often termed as *robustness.*

### 1.1 Preliminaries

Let the life length of a device, say $X$, have density function $g(x;\theta) = \theta^{-1}$ exp $(x/\theta)$, where $\theta > 0$ is the mean time to failure.

In MIL-STD-781C we test the simple hypothesis $H_0{:}\theta = \theta_0$, versus the alternative hypothesis $H_1 : \theta = \theta_1$, where $0 < \theta_1 < \theta_0 < \infty$. The probability that $H_0$ is rejected when $\theta = \theta_0$ is known as the *producer's risk*, and is denoted by $\alpha$, whereas the probability that $H_0$ is accepted when $\theta = \theta_1$ is known as the *consumer's risk*, and is denoted by $\beta$. It is customary to refer to $\theta_0$ as the *desired mean*, and to $\theta_1$ *as the minimum acceptable mean;* the quantity $(\theta_0/\theta_1) \overset{\text{def}}{=} d$ is referred to as the *discrimination ratio.* The risks $\alpha$ and $\beta$ take values between .01 and .20, whereas $d$ is between 1.5 and 3.0.

In MIL-STD-781C, it is possible for us to test either one item at a time or several items at a time. In our analyses the two cases will be studied separately.

Suppose then, that $n \geqq 1$ items are put on the test rack, and their ordered times to failure, say $X_{(1)} \leqq X_{(2)} \leqq \ldots \leqq X_{(r)}$, noted. Failed items could either be replaced or not; in the latter case, $r \leqq n$. For any $t \in [0, \infty)$, the quantity

$$V(t) \overset{\text{df}}{=} \begin{cases} nt, & \text{if a failed item is replaced;} \\ \sum_{i=1}^{r} X_{(i)} + (n-r)(t - X_{(r)}) & \text{if a failed item is not replaced;} \end{cases} \tag{1.1}$$

is known as the *total time on test,* and it indicates the total lifetime observed on the test rack.

Information about $\theta$ is continuously available through $V(t)$, and by a direct application of Wald's sequential probability ratio test, we let $A = (1-\beta)/\alpha$, $B = \beta/(1-\alpha)$, and do one of the following:

$$\begin{aligned} &\text{accept} \quad H_0 \quad \text{if } (\theta_0/\theta_1)^r \exp\left[-\left(\frac{1}{\theta_0} - \frac{1}{\theta_0}\right) V(t)\right] \leqq B, \\ &\text{reject} \quad H_0 \quad \text{if } (\theta_0/\theta_1)^r \exp\left[-\left(\frac{1}{\theta_1} - \frac{1}{\theta_0}\right) V(t)\right] \geqq A; \end{aligned} \tag{1.2}$$

continue testing if either one of the above is not satisfied.

In the nonreplacement case, if no decision has been made by the time $t = X_{(n)}$, we put $n$ more fresh items on the test rack and continue with the procedure described above.

Let $L_G(\theta)$ denote the probability that $H_0$ is accepted when the mean time to failure is $\theta$; the subscript $G$ associated with $L$ is to denote the fact that the underlying distribution of $X$ is an exponential distribution. It is obvious that $L_G(\theta_0) \simeq 1 - \alpha$ and that $L_G(\theta_1) \simeq \beta$; it can be shown that $L_G(\theta = S) \simeq \log A/(\log A - \log B)$, where $S = \log(\theta_0/\theta_1)/((1/\theta_1) - (1/\theta_0))$. Recall that a plot of $L_G(\theta)$ versus $\theta$ is known as the *operating characteristic curve* (O-C curve) of the procedure. Let $E_{G_\theta}(r)$ denote the expected number of failures to reach a decision when $\theta$ is the mean time to failure, and the underlying distribution of lifetimes is $G$.

*Definition 1.1:* If a distribution function, say $F$, with a density function $f$ has a *failure rate* $r(x) \overset{\text{df}}{=} f(x)/(1-F(x))$, which is nondecreasing (nonincreasing) in $x$, for all $x \geqq 0$, and $F(x) \not\equiv 1$, then $F$ is said to be IFR (DFR).

The abbreviation IFR (DFR) is used to denote increasing (decreasing) failure rate. Most of the commonly use life distributions, such as the gamma and the Weibull, are either IFR or DFR. By convention, an exponential distribution is both IFR and DFR. Distributions which are either IFR or DFR are also known as "distributions having a monotone failure rate."

## 1.2 Investigation of Robustness

Suppose that the lifetimes $X$ now have a distribution function $F$, where $F$ need not be exponential. Suppose that $E_F(X) = \theta$; that is, $F$ has mean $\theta$. Also, let $H_0$ and $H_1$ denote the same hypotheses as described in Section 1.1.

We shall first consider the case wherein only one item is on the test rack at any point in time; that is, $n = 1$. For any $t$, $t \in [0, \infty)$, we compute $V(t)$ using (1.1) and obey the decision rule specified in (1.2), except that in order simplify our analyses we adopt the convention that a decision to accept or reject a hypothesis can only be made at the time of failure, and not between failures.

If we let $L_G(\theta)$ denote the probability that $H_0$ is accepted when $\theta$ is the mean time to failure, then we can prove [see Montagne and Singpurwalla (1982)]

*Theorem 1.2:* If $F$ is IFR (DFR) with mean $\theta$, then

$$L_G(\theta)\begin{cases} \leqq (\geqq) L_G(\theta), & \theta < S \\ = L_G(\theta), & \theta = S \\ \geqq (\leqq) L_G(\theta), & \theta > S. \end{cases}$$

Theorem 1.2 is important because it establishes a relationship between the O-C curves for the sequential procedure described by (1.2) when the underlying distribution of lifetimes is an exponential with mean $\theta$ or an IFR (DFR) distribution also with mean $\theta$. As a corollary to Theorem 1.2, we have

*Corollary 1.3:* Under the conditions of Theorem 1.2,

$$\alpha_I \leqq \alpha \leqq \alpha_D,$$

and

$$\beta_I \leqq \beta \leqq \beta_D,$$

where $\alpha_I(\alpha_D)$ and $\beta_I(\beta_D)$ denote the consumer's and the producer's risks, respectively, when $F$ is IFR (DFR).

The practical implication of Corollary 1.3 is that when the sequential part MIL-STD-781C procedure is used for distributions which are not exponential, *both* the producer's and the consumer's risks are lower (higher) than their specified values, provided that the underlying distribution is IFR (DFR). This result confirms and generalizes the empirical results of Harter and Moore (1976) for the special case of the Weibull distribution. As is to be expected, the reduction in risks is compensated by an increase in the expected time to finish testing. To be more specific, let $E_{F_\theta}(r)$ denote the expected number of failures to reach a decision when the underlying distribution of lifetimes is $F$ with mean $\theta$, and the procedure of (1.2) is used. Then, we can prove

*Theorem 1.4:* When $F$ is IFR (DFR) with mean $\theta$

$$E_{F_\theta}(r) \underset{(\leqq)}{\geqq} E_{G_\theta}(r).$$

In view of Theorem 1.4, we question the advantage of using the sequential part of MIL-STD-781C when the underlying distribution of life lengths is not exponential. Under such circumstances, it is perhaps more economical to use the fixed sample size test procedures described in MIL-STD-781C.

Bounds on $E_{F_\theta}(r)$ have also been obtained by Montagne and Singpurwalla (1982). These, however, are in terms of $L_G(\theta)$, and are as follows:

$$\frac{h_1 - L_G(\theta)(h_0 + h_1)}{S - \theta} \leqq E_{F_\theta}(r) \leqq \begin{cases} \dfrac{\log A}{S - \theta}, & \theta < S \\ \dfrac{\log B}{S - \theta}, & \theta > S, \end{cases}$$

and

$$\frac{h_0 h_1}{S^2} \leqq E_{F_\theta}(r) \leqq \infty, \quad \theta = S,$$

where

$$h_0 = -\frac{\log B}{\left(\dfrac{1}{\theta_1} - \dfrac{1}{\theta_0}\right)} \text{ and } h_1 = -\frac{\log A}{\left(\dfrac{1}{\theta_1} - \dfrac{1}{\theta_0}\right)}.$$

Our results for the case when multiple items ($n > 1$) are put on the test rack are in marked contrast to those given in Theorem 1.2 and Corollary 1.3.

To see this, suppose that $n > 1$ items are put on the test rack, and the procedure described in (1.2) is used; failed items are not replaced. A common procedure is to truncate the test after $n_0$ failures, where $n_0 \leqq n$. This ensures that the sequential testing procedure will not continue for an indefinite period of time. When the test is so terminated, an arbitrary decision is made either to accept or to reject $H_0$. Theorem 1.5 given below is proved using the fact that the testing procedure is terminated after $n_0$ failures have been observed.

Let $L_{F,n}(\theta)$ be the probability that $H_0$ is accepted when $n$ items are put on the test rack; there is no replacement of failed items, and the underlying distribution of lifetimes is $F$ with mean $\theta$. Then we can prove that as long as the failure rate of $F$ is zero at time 0, the following is true.

*Theorem 1.5:* If $F$ is IFR (DFR) with mean $\theta$

$$L_{F,n}(\theta) \uparrow (\downarrow)\, n, \quad \forall\ \theta.$$

The practical implication of Theorem 1.5 is that a *devious producer,* who is aware of the fact that his items have an IFR life distribution with mean $\theta <<$ $\theta_1$, can *increase his chances* of having his lot accepted by placing a large number of items on test at one time (i.e., making $n$ large). Similarly, a *vindictive* consumer can *lower the chances* of acceptance when his items have a DFR life distribution with $\theta >> \theta_0$.

We hope that the conclusions of Theorems 1.2 and 1.5 will be incorporated into a revision of the Military Standard 781C.

## 2. A MODEL FOR STEP-STRESS TESTING

Environmental or accelerated testing is an important feature of MIL-STD-781C. The goals of environmental testing are two-fold:

(i) to weed out items that are weak or defective by inducing early failures due to increased environmental (stress) conditions, and
(ii) to infer about the lifetimes of the items under use conditions environment by observing their lifetimes under increased environmental conditions.

In our research we have focused attention on the second of these two goals.

Most of the published statistical literature on accelerated life testing deals with the situation wherein items are subjected to larger than normal values of the environmental stresses, which are held fixed throughout the life test. Such tests are known as *fixed-stress* tests. In *step-stress* testing, the stress on an unfailed item is allowed to change at preassigned test times to successively higher and higher values, thereby increasing the chances of failure of an unfailed item.

Specifically, let $V_0 < V_1 < \ldots < V_I$ be a collection of accelerated stresses, with $V_0$ the use condition stress. At the start of the test, an item is subjected to a stress $V_{i_1}$ for a time period $[0, t_1)$, where $i_1 \in (0, 1, \ldots, I)$, and $t_1 < \infty$. If the item has not failed at time $t_1$, the stress is increased to $V_{i_2}$, where $i_2 \in (i_1 + 1, \ldots, I)$, and this procedure is continued until time $t_J$. The time points $t_1, t_2, \ldots, t_J$ are preassigned. Schematically, we have

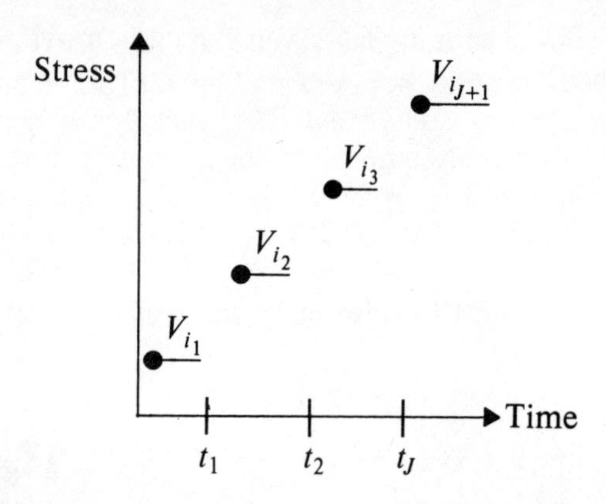


The collection $[(V_{i_1}, t_1), (V_{i_2}, t_2), \ldots, (V_{i_J}, t_J), V_{i_{J+1}}]$ is called the *stress pattern* and is denoted by Shaked and Singpurwalla (1982) by $[\underset{\sim}{V} \square \underset{\sim}{t}]$, where $\underset{\sim}{V} = (V_{i_1}, V_{i_2}, \ldots, V_{i_{J+1}})$ and $\underset{\sim}{t} = (t_1, t_2, \ldots, t_J)$.

## 2.1 The Model

Let $F_i$ denote the distribution of life lengths under $V_i$, and let

$$F_i(t) = F(CV_i^\alpha t) \tag{2.1}$$

where $F$ is some unknown distribution function; $C$ and $\alpha$ are unknown constants. The relationship (2.1) is the well known *power law*, commonly used in accelerated testing.

Let $T$ be the time to failure of a device under the stress pattern $(\underset{\sim}{V} \square \underset{\sim}{t})$, and let $H$ be the distribution function of $T$. Thus,

$$\begin{aligned} H(t) = P(T \leqq t) &= F_{i_1}(t) = F(CV_i^\alpha t), & t \in [0, t_1), \\ &= F(CV_{i_1}^\alpha t + CV_{i_2}^\alpha (t - t_1)), & t \in [t_1, t_2) \\ &= F\left[C \sum_{j=1}^{l} V_{i_j}^\alpha (t_j - t_{j-1}) + CV_{i_{l+1}}^\alpha (t - t_l)\right], & t \in [t_l, t_{l+1}), \end{aligned} \tag{2.2}$$

for $l = 0, 1, \ldots, J$; $t_0 \equiv 0$ and $t_{J+1} = \infty$.

Before we give a justification for this model, we note that when $J = 1$ and $F$ is the exponential distribution, (2.2) becomes the model considered by De Groot and Goel (1979). We can also show that under some special cases, (2.2) is the model considered by Nelson (1980). These authors give little motivation for their models.

## 2.2 Motivation of the Model

There are two different scenarios that can be used to motivate our model.

As a first scenario, assume that under $V_i$, shocks occur according to the postulates of a Poisson process with a rate $\lambda_i = CV_i^{\alpha}$, $i = 1, \ldots, I$. Then

$$\bar{H}_i(t) = P\{T \geqq t\} = \sum_{m=0}^{\infty} \frac{e^{-\lambda_i t}(\lambda_i t)^m}{m!} \bar{P}_m, \; t \geqq 0,$$

where $\bar{P}_m$ is the probability that the item survives $m$ shocks. Clearly, for $t \in [t_l, t_{l+1})$, $l = 0, 1, \ldots, J$,

$$\bar{H}(t) = \sum_{m=0}^{\infty} \exp\left\{-\left[\sum_{j=1}^{l} \lambda_{i_j}(t_j - t_{j-1}) + \lambda_{i_{l+1}}(t - t_l)\right]\right\}$$
$$\times \left[\sum_{j=1}^{l} \lambda_{i_j}(t_j - t_{j-1}) + \lambda_{i_{l+1}}(t - t_l)\right]^m \frac{\bar{P}_m}{m!}.$$

We note that our model (2.2) holds if we take

$$\bar{F}_i(t) = 1 - F_i(t) = \sum_{m=0}^{\infty} \frac{e^{-\lambda_i t}(\lambda_i t)^m}{m!} \bar{P}_m.$$

As a second scenario, we consider what is known as the concept of "damage accumulation." We assume that under $V_i$, damage accumulates at a rate $\lambda_i = CV_i^{\alpha}$. Let $X$ be a threshold which the item can sustain. That is, the item fails if the damage exceeds $X$. Suppose that $X$ is a random variable with distribution function $F$. Thus under $V_i$, $F_i(t) = P(CV_i^{\alpha} t > X) = F(CV_i^{\alpha} t)$, and consequently under $(\underset{\sim}{V} \square \underset{\sim}{t})$, (2.2) results.

There are two remarks about our model:

(i) the model reflects the fact than an item is not like new when a change in stress occurs, and
(ii) since $\bar{P}_m$ is not assumed, $F$ is not specified, and so our model is nonparametric in this sense.

## 2.3 Estimation and Inference

The estimation of $F_0$, the life distribution under use conditions stress, is undertaken by first estimating $\alpha$ using failure data obtained under the step-stress test, and then transforming all the observed lifetimes to correspond to lifetimes under $F_0$. The transformed lifetimes are then used to obtain $\hat{F}_0$, the empirical distribution function of $F_0$. The details of this procedure and the resulting formulae are given in Shaked and Singpurwalla (1982).

# 3. A BINARY RESPONSE DAMAGE PREDICTION PROBLEM

The problem that we describe below has arisen out of some consulting work that we have been doing for the Explosions Effects Branch of the Naval Surface

Weapons Center at White Oak, Maryland on behalf of the Office of Naval Research, and the Ballistics Research Laboratory at Aberdeen Proving Grounds Maryland, on behalf of the Army Research Office. Here, target models of a submarine hull are tested for their ability to withstand damage induced by underwater nuclear explosion shocks.

Accordingly, if $0 \equiv S_0 < S_1 < \ldots < S_i < \ldots < S_M < S_\infty$ denote the shock intensities, and if only *one* model is tested under each shock intensity, then we observe a binary random variable $X_i$, where

$$X_i = 1, \text{ if the model is damaged under stress } S_i, \text{ and}$$
$$= 0, \text{ if the model is not damaged; } i = 0, 1, \ldots, M.$$

Let $P\{X_i = 1\} = p_i$, $i = 0, 1, \ldots, M$, and let us assume that for any $j \in [0, \infty)$, $p_j$ is *convex* in $j$, for $j = 0, 1, \ldots, M$; furthermore, we set $p_0 \equiv 0$, and $p_\infty \equiv 1$.

Given $X_1, X_2, \ldots, X_M$, our aim is to obtain estimates $\hat{p}_1, \ldots, \hat{p}_M$ such that $\hat{p}_i$ is convex in $i$, and for any $j \neq 1, 2, \ldots, M$, we need to obtain an estimate of $p_j$, say $\hat{p}_j$.

The stated problem also occurs in several other contexts, such as bioassay (where typically several items are tested at each dosage), reliability growth testing (where an increase in the shock intensity corresponds to an increase in time), and sensitivity testing experiments performed by those interested in studing the strength of materials. The assumption of convexity is specific to our problem and is the reason for the novelty of our approach. The approach that we take here is Bayesian, and is an extension of that taken by Ramsey (1972) in the context of bioassay; the convexity motivates the extension.

To simplify our analyses, we let $S_i - S_{i-1} = \Delta$, $i = 1, \ldots, M$, and let $Z_i = p_i - p_{i-1}$. Even though the discussion here assumes equispaced shock intensities, this requirement is not essential. Let $Y_i = Z_i - Z_{i-1}$, $i = 1, \ldots, M$ be the second differences of the $p_i$'s, and because of convexity

$$Z_1 \leqq Z_2 \leqq \cdots \leqq Z_M \text{ and } \sum_{i=1}^{M} Z_i \leqq 1.$$

Further, the $Y_i$'s are greater than or equal to 0, with $Z_1 = Y_1$, $Z_2 = Y_2 + Y_1$, $\ldots$, $Z_M = Y_1 + Y_2 + \ldots + Y_M$. Also, since $\sum_{i=1}^{M} Z_i \leqq 1$, it follows that $U_1 + U_2 + \ldots + U_i + \ldots + U_M \leqq 1$, where $U_i = (M - i+1)Y_i$, $i = 1, \ldots, M$. With some algebra we can also verify that

$$U_1 = MP_1,\ U_2 = (M-1)(p_2 - p_1),\ \ldots,\ U_i = (M-i+1)(p_i - 2p_{i-1} + p_{i-2}),$$

and $U_M = p_M - 2p_{M-1} + p_{M-2}$.

Since $U_i \geqq 1$, and $\sum_{i=1}^{M} U_i \leqq 1$, a suitable prior distribution for the $U_i$'s is a Dirichlet distribution, with density function [cf. Ramsey (1972)]

$$f(U_1, \ldots, U_M) =$$
$$\frac{\Gamma(\beta\alpha_1 + \ldots + \beta\alpha_{M+1})}{\Gamma(\beta\alpha_1) \ldots (\Gamma(\beta\alpha_{M+1})} U_1^{\beta\alpha_1 - 1} \ldots U_M^{\beta\alpha_M - 1}(1 - U_1 - \ldots - U_M)^{\beta\alpha_{M+1} - 1},$$

where $\beta$ and $\alpha_j$, $j = 1, \ldots, M + 1$, are prior parameters with the property that $\sum_{j=1}^{M+1} \alpha_j = 1$, and $\beta > 0$.

It is easy to verify that the marginal probability density function of $U_i$ is a beta density on (0, 1), with parameters $\beta\alpha_i$ and $\beta(1 - \alpha_i)$. Thus,

$$E(U_i) = \alpha_i = E[(M - i + 1)(p_i - 2p_{i-1} + p_{1-2})].$$

Let $p_i^*$ be the best guess of $p_i$, $i = 1, \ldots, M$. Then, the prior parameters $\alpha_i$ can be obtained by setting $\alpha_1 = Mp_i^*$, and $\alpha_i = (M - i + 1)(p_i^* - 2p_{i-1}^* + p_{i-2}^*)$, $i = 1, \ldots, M$. The choice of the parameter $\beta$ is more complicated; $\beta$ controls the probability that $p_i$ is close to $p_{i-1}$ or to $p_{i+1}$. For a more detailed discussion of the choice of $\beta$, we refer the reader to Ramsey (1972), or to Mazzuchi and Singpurwalla (1982).

The Bayes estimator of $p_1, \ldots, p_M$ is obtained by finding those values $\hat{p}_1, \ldots, \hat{p}_M$ which maximize the posterior density

$$\prod_{i=1}^{M} p_i^{X_i}(1 - p_i)^{1-X_i}\{(Mp_1)^{\beta\alpha_1 - 1}[(M-1)(p_2 - 2p_1)]^{\beta\alpha_2 - 1} \ldots (1 - p_M)^{\beta\alpha_{M+1} - 1}\}$$

subject to the restriction that

$$2p_{i-1} - p_{i-2} \leqq p_i \leqq 1 + \frac{M - i}{M - i + 1} p_{i-1}, \; i = 1, \ldots, M.$$

This constrained maximization is performed using a special code in nonlinear optimization. The constraint (restriction) mentioned above is brought about by the assumption of convexity. The technical details supporting the above approach will be given in a forthcoming report by Habibullah, Shaked, and Singpurwalla (1982). A user's guide describing the optimization code is to appear in a forthcoming report by Mazzuchi and Soyer (1982).

## 4. THE STOCHASTIC CHARACTERIZATION OF LIFE LENGTHS GENERATED BY A MINIMAL REPAIR POLICY

Much of the published literature dealing with the stochastic behavior of life lengths of repaired items assumes that a failed item is restored to the status of a brand new item as a result of repair. This assumption is reasonable when repair involves replacement or a complete overhaul. In practice, repair is often imperfect; a failed item is often restored to a condition which is either slightly better than its condition prior to failure, or to a condition *equal* to its condition prior to failure. Repair actions describing this latter policy are termed "minimal repair" by Barlow and Hunter (1960) and "bad as old" by Ascher and Feingold

(1969). The material given below is based upon Balaban and Singpurwalla (1981), who attempt to formalize the stochastic behavior of life lengths generated by minimal repair actions, and give some of their properties.

Let $X(n)$ denote the time to the $n$th failure of a device which is functioning at time 0. We let $X(0) \equiv 0$, and note that $X(1) \leqq X(2) \leqq \ldots \leqq X(n) \leqq X(n+1) \leqq \ldots$.

*Definition 4.1:* The stochastic process $\{X(n);\ n = 0, 1, \ldots\}$ is said to be *age persistent* (AP) if and only if, for $0 \leqq x < y < \infty$,

$$P\{X(n+1) \geqq y \mid X(n) = x,\ X(n-1) = \cdot,\ \ldots\} = P\{X(1) \geqq y \mid X(1) \geqq x\}.$$

The above definition states that the distribution of life lengths following the $n$th failure, given that the $n$th failure occurred at $x$, is the same as the distribution of the first life length given that it was at least $x$. Intuitively, this characterizes minimal repair, since here failure and the associated maintenance action has no discernible effect on the aging process.

In contrast to the above, we have perfect repair described by

*Definition 4.2:* The stochastic process $\{X(n);\ n = 0, 1, \ldots\}$ is said to be *age independent* (AI) if and only if

$$P\{X(n+1) \geqq y \mid X(n) = x,\ X(n-1) = \cdot,\ \ldots\} = P\{X(1) \geqq y - x\},$$

for all $0 \leqq x < y < \infty$.

We note that an ordinary renewal process is an age independent process. Writing $\bar{F}_X(x)$ to denote $P\{X \geqq x\}$, we thus have

$$\bar{F}_{X(n+1)|X(n)}(y \mid x) = \begin{cases} \bar{F}_{X(1)}(y) \mid \bar{F}_{X(1)}(x), & \text{under (AP)} \\ \bar{F}_{X(1)}(y - x), & \text{under (AI)} \end{cases} . \tag{4.1}$$

Properties of life distributions which are of interest in reliability theory are given in Barlow and Proschan (1975, Ch. 4). Typical of these are those which characterize aging or wearout, such as increasing (decreasing) failure rate IFR (DFR), increasing (decreasing) failure rate average IFRA (DFRA), new better (worse) than used NBU (NWU), new better (worse) than used in expectation NBUE (NWUE), and increasing (decreasing) mean residual life IMRL (DMRL). These properties are well known, and are also of interest when studying life lengths under repair. To conserve space, we refer the reader to Chapter 4 of Barlow and Proschan (1975) for defining characteristics of these properties.

The following results are typical of those given by Balaban and Singpurwalla (1981), and may be of both practical and theoretical interest.

*Theorem 4.3:* If the stochastic process $\{X(n);\ n = 0, 1, \ldots\}$ is both (AP) and (AI), then $\bar{F}_{X(1)}$ must be an exponential distribution.

The above theorem establishes the role of the exponential distribution in a study of the quality of repair policies.

In the following theorem, we describe the stochastic behavior of the time interval between successive repair times, given the ageing characteristics of a new item, under age persistence and age independence. To see this, we first need to define

$$T(i+1) = X(i+1) - X(i), \quad i = 0, 1, 2, \ldots .$$

*Theorem 4.4:* If $\bar{F}_{X(1)}$ is IFR (DFR), then

$$P\{T(i+1) > t \mid X(i) = x, \text{ and (AP)}\} \leqq (\geqq)\, P\{T(i+1) > t \mid X(i) = x, \text{ and (AI)}\}.$$

By letting $N(x, y)$ denote the number of failure/repair actions in the time interval $(x, y)$, $r(u) = dF_{X(1)}(u)/\bar{F}_{X(1)}(u)$ denote the *failure rate* of $\bar{F}_{X(1)}$ at any $u \geqq 0$, and $R(t) = \int_0^t r(u)\,du$, then the counting process under age persistence is given by

*Theorem 4.5:* If $\{X(n);\ n = 0, 1, \ldots\}$ is (AP),

$$P\{N(x, y) = m\} = e^{-[R(y) - R(x)]}[R(y) - R(x)]^m / m!,$$

for $m = 0, 1, 2, \ldots$ .

Thus the counting process under age persistence is a nonhomogeneous Poisson process with an intensity function as described above.

Bounds on the conditional distribution of $X(n+1)$ given $X(n) = x$ under age persistence are given in

*Theorem 4.6:* If $\{X(n);\ n = 0, 1, \ldots\}$ is (AP), and if $\bar{F}_{X(1)}$ is IFRA (DFRA), then

$$\bar{F}_{X(n+1)|X(n)}(y \mid x) \underset{(\geqq)}{\leqq} [\bar{F}_{X(1)}(x+y)]^{(y-x)/y}.$$

The following closure theorem, though not of much practical use, is of interest from the point of view of further research. To state this, let $S$ denote a class of distributions, and we shall need

*Definition 4.7:* $S$ is said to be *closed under age persistence*, denoted by $S$ is $C_{(AP)}$, if

$$\bar{F}_{X(1)} \in S \Longrightarrow \bar{F}_{X(n+1)|X(n)} \in S, \text{ for all } n.$$

Nonclosure under age persistence is denoted by $S$ is $\bar{C}_{(AP)}$. We now have

*Theorem 4.8:* (i) IFR (DFR) is $C_{(AP)}$,

(ii) DMRL (IMRL) is $C_{(AP)}$,

(iii) IFRA, NBU, NBUE, and their analogues are $\overline{C}_{(AP)}$.

## REFERENCES

H. Ascher and H. Feingold (1969), "Bad as old" analysis of system failure, *Ann. Assurance Sci., 8th R&M Converence*, Gordon and Breach, New York, 49-62.

H. Balaban and N.D. Singpurwalla (1981), The stochastic characterization of a sequence of life lengths under minimal repair, Technical Paper Serial T-443, Institute for Management Science and Engineering, The George Washington University. Revised Feb. 82.

R. Barlow, and M. Hunter (1960), Optimum preventive maintenance policies, *Operations Res.* **8**, 90-100.

R. Barlow and F. Proschan (1975), *Statistical Theory of Reliability and Life Testing*, Holt, Rinehart, and Winston, New York.

M. DeGroot and P. Goel (1979), Bayesian estimation and optimal designs in partially accelerated life testing, *Naval Res. Logist. Quart.* **26**, 223-235.

B. Epstein and M. Sobel (1955), Sequential life tests in the exponential case, *Ann. Math. Statist.* **28**, 82-93.

M. Habibullah, M. Shaked and N.D. Singpurwalla (1982), A binary response damage estimation and prediction model. In preparation.

L. Harter and A. Moore (1976), An evaluation of exponential and Weibull test plans, *IEEE Trans. in Reliability* **R-25**, 100-104.

E. Montagne and N.D. Singpurwalla (1982), On the robustness of the sequential exponential life testing procedures. In preparation.

T.A. Mazzuchi and R. Soyer (1982), A user's guide to a computer programs for a binary response damage estimation and prediction model, Technical Report GWU/IRRA/TR-82/1, The George Washington University, Washington, D.C.

W. Nelson (1980), Accelerated life testing—step stress models and data analysis, *IEE Trans. in Reliability, R-29* (2), 103-108.

G. Ramsey (1972), A Bayesian approach to bioassay, *Biometrics* **28**, 841-858.

M. Shaked and N.D. Singpurwalla (1981), Inference for step-stress accelerated life testing. To appear, *Journal of Statistical Inference and Planning.*

T.A. Mazzuchi and N.D. Singpurwalla (1982), The U.S. Army (BRL's) Kinetic Energy Penetrator Problem: Estimating the Probability of Response for a Given Stimulus, *Proceedings of the 1981 Army Design of Experiments Meeting*, (to appear).

# Some Reliability Concepts Useful for Materials Testing

Richard A. Johnson and G. K. Bhattacharyya

*University of Wisconsin, Madison*

**Abstract**

Stress and strength analyses are of widespread practical interest in materials testing and acquisition decisions. Because equipment failure is a rare occurrence, due to the happy consequence of technological progress, the gathering of a comprehensive set of failure data is increasingly expensive and time consuming. The success of our analyses depend crucially on the proper choice of statistical models. The aspect of model selection is examined in light of a few life test experiments. Some of our recent advances on stress-strength analyses are outlined. These include prevalent situations where some measurable covariates affect the strength or the stress.

## 1. INTRODUCTION

Decisions involved in the acquisition process of a material must be guided by available quantitative informations concerning the strength, durability or other appropriate measure of quality, as well as the severity of the stress conditions under which the material is intended to be deployed. Reliability engineers are naturally confronted with the design, analysis and interpretation of life testing or strength testing of materials either under controlled loadings or stochastic environmental stresses. In this article, we present an outline of two aspects of our research that have particular relevance to the modeling and analysis of these experiments.

We first address the issue of selecting a statistical distribution that describes the variation in a material's property like the tensile strength. In Section 2, we describe a practically useful procedure for choosing between the three widely used parametric families of distributions, namely, the normal, lognormal, and the three-parameter Weibull family. The choice is based on

Research supported by Office of Naval Research Grant No. N00014-78-C-0722.

computer-aided maximum likelihood estimation and plots of the order statistics versus their expected or median values. The proposed method is illustrated with a data set of compression strength of a glass composite. The fitted Weibull model is employed to construct confidence intervals for a percentile.

Some of our recent advances on the analysis of stress versus strength model of reliability are outlined in Section 3. These concern situations where some observable concomitants influence the strength and/or the stress variables, and should therefore be incorporated into the model. We have considered Bayesian estimation of reliability in a stress-strength setting under the assumptions of normal distributions and linear regression of the covariates.

## 2. SELECTING A DISTRIBUTION

Given data on the strength of a material (see Fig. 1) or the lifetime of components, we ask: Can the data be adequately described by a common parametric distribution? In the context of reliability applications, the important candidates are

$$\text{Normal:} \quad f(x) = \frac{1}{\sqrt{2\pi}\,\sigma} e^{-\frac{1}{2\sigma^2}(x-\mu)^2}, \quad -\infty < x < \infty$$

$$\text{Lognormal:} \quad f(x) = \frac{1}{\sqrt{2\pi}\,\eta x} e^{-\frac{1}{2\eta^2}(\ln x-\delta)^2}, \quad 0 < x < \infty$$

$$\text{Three-parameter Weibull:} \quad f(x) = \frac{p}{\theta}\left(\frac{x-\beta}{\theta}\right)^{p-1} e^{-\left(\frac{x-\beta}{\theta}\right)^p}, \quad \beta < x < \infty$$

The procedure we discuss can be extended to include others.

Once a distribution is selected, one typically wants, from a quality control standpoint, to obtain a point estimate of a specified lower percentile of the distribution or even a lower confidence bound.

Because lack of fit may occur in an unlimited number of ways, no hard and fast rules can be set for the selection process. We have, however, found the following procedure very helpful in our choice of a distribution.

First plot the order statistics $x_{(1)} \leqslant \cdots \leqslant x_n$ versus their (approximate) expected values. The normal plot of the $x_{(i)}$ and another for $\ln x_{(i)}$ are standard in many software packages. For the Weibull alternative, we plot $x_{(i)}$ versus $\left[\ln\left(1 - \frac{i-3/8}{n+1/4}\right)\right]^{1/\hat{p}} \hat{\theta} + \hat{\beta}$, or just $\left[\ln\left(1 - \frac{i-3/8}{n+1/4}\right)\right]^{1/\hat{p}}$, where the former relates to median of the $i^{\text{th}}$ order statistic. Conformity of the data to a straight line pattern can be inspected visually. Figures 2(a), (b) and (c) show these plots for the data of Fig. 1.

The correlation coefficient, for the normal plot, can be used to test for the goodness-of-fit. An approximate 10% critical value, for $n = 48$, is .981. Thus, both the normal and lognormal pass this screen. The lognormal case appears to have the greatest deviation from a straight line fit.

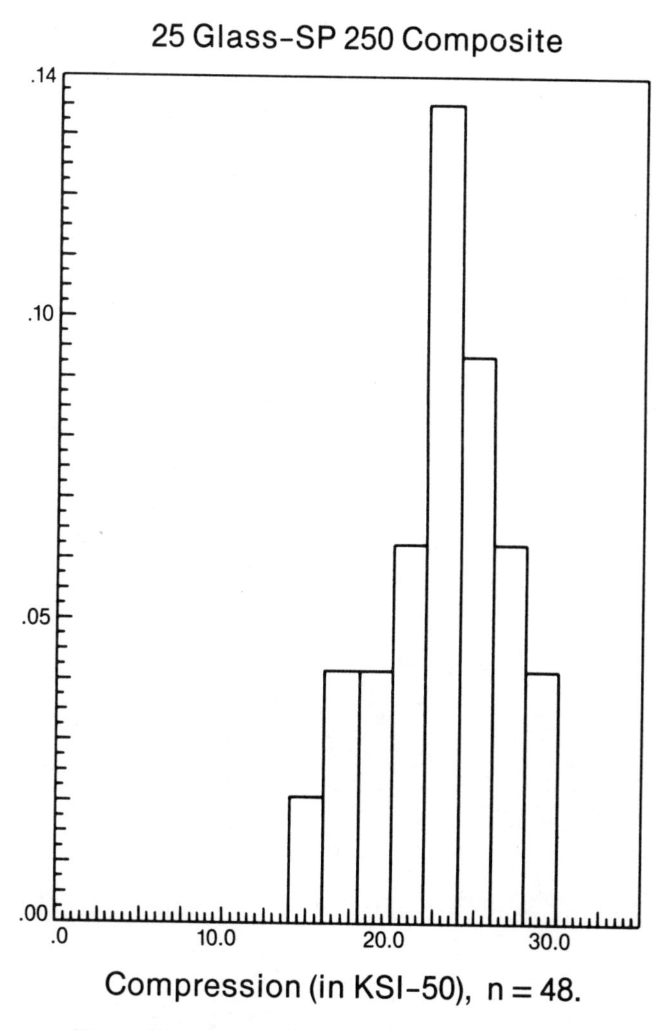


**Fig. 1. Histogram of the compression strength of glass.**

The distribution of correlation coefficient, determined from the Weibull plot, is still unknown. Monte Carlo studies show that it does depend on the true shape parameter for the range of sample sizes 75-200 considered. Our approximate 10% critical curve, depending on $\hat{p}$, suggests that the observed correlation does not contradict the assumption of the Weibull population.

We are currently working on extensions of the goodness-of-fit procedure to include censored observations.

After completing the goodness-of-fit phase, distributions that are still viable are compared on the basis of the maximized likelihoods.

$$\text{Normal:} \quad \hat{L}_N = \prod_{i=1}^{n} \frac{1}{\sqrt{2\pi}\hat{\sigma}} e^{-\frac{1}{2}\left(\frac{x_i-\hat{\mu}}{\hat{\sigma}}\right)^2} = \frac{1}{(2\pi\hat{\sigma}^2)^{n/2}} e^{-n/2}$$

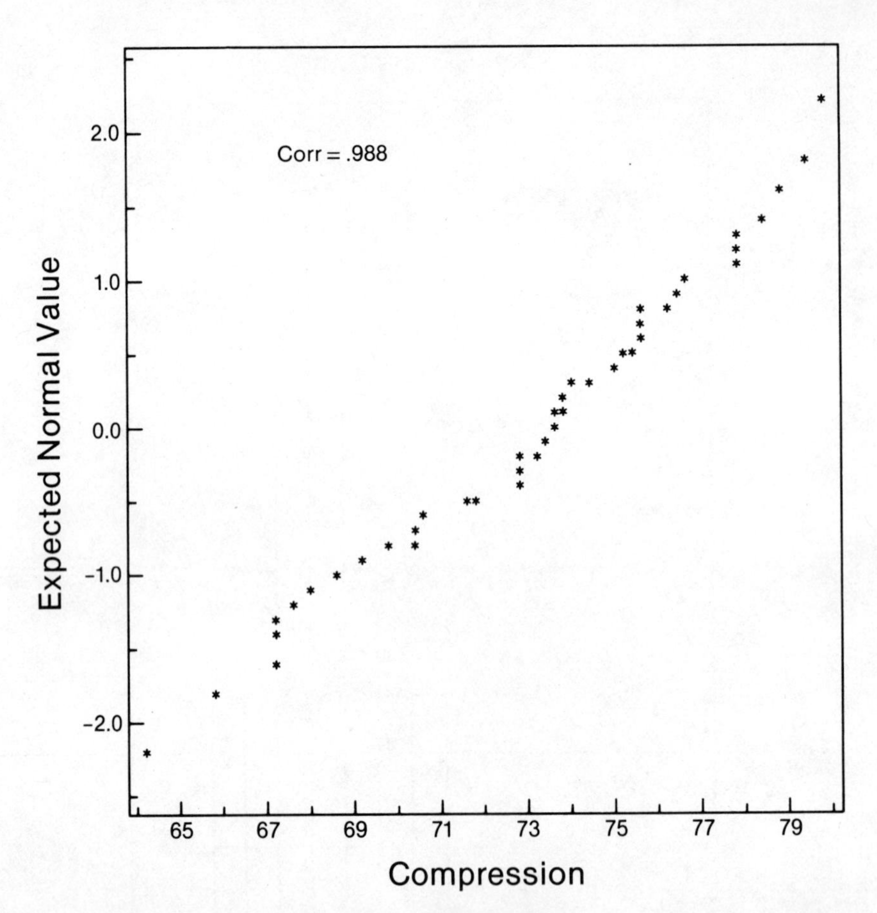


**Fig. 2(a). Normal Plot**

$$\text{Lognormal:} \quad \hat{L}_{LN} = \prod_{i=1}^{n} \frac{1}{\sqrt{2\pi}\hat{\eta} x_i} e^{-\frac{1}{2}\left(\frac{\ln x_i - \hat{\delta}}{\hat{\eta}}\right)^2} = \frac{1}{(2\pi\hat{\eta}^2)^{n/2}} e^{-n/2} \frac{1}{\prod_{i=1}^{n} x_i}$$

$$\text{Weibull:} \quad \hat{L}_W = \prod_{i=1}^{n} \frac{\hat{p}}{\hat{\theta}} \left(\frac{x_i - \hat{\beta}}{\hat{\theta}}\right)^{\hat{p}-1} e^{-\left(\frac{x_i - \hat{\beta}}{\hat{\theta}}\right)^{\hat{p}}}$$

The maximum likelihood estimates $\hat{p}$, $\hat{\theta}$, $\hat{\beta}$ for the Weibull parameters are obtained by a computer search routine.

From the 48 measurements on the compression strength of the glass composite, we obtain $\hat{p} = 6.098$. $\hat{\theta} = 20.507$ and $\hat{\beta} = 54.088$.
Further, $\hat{\sigma}^2 = 13.623$ so

$$\frac{1}{n} \ln \hat{L}_N = -2.725, \ \frac{1}{n} \ln \hat{L}_{LN} = -2.735, \ \frac{1}{n} \ln \hat{L}_W = -2.703$$

and, on this basis, we prefer the Weibull. Figure 3 shows the fitted Weibull

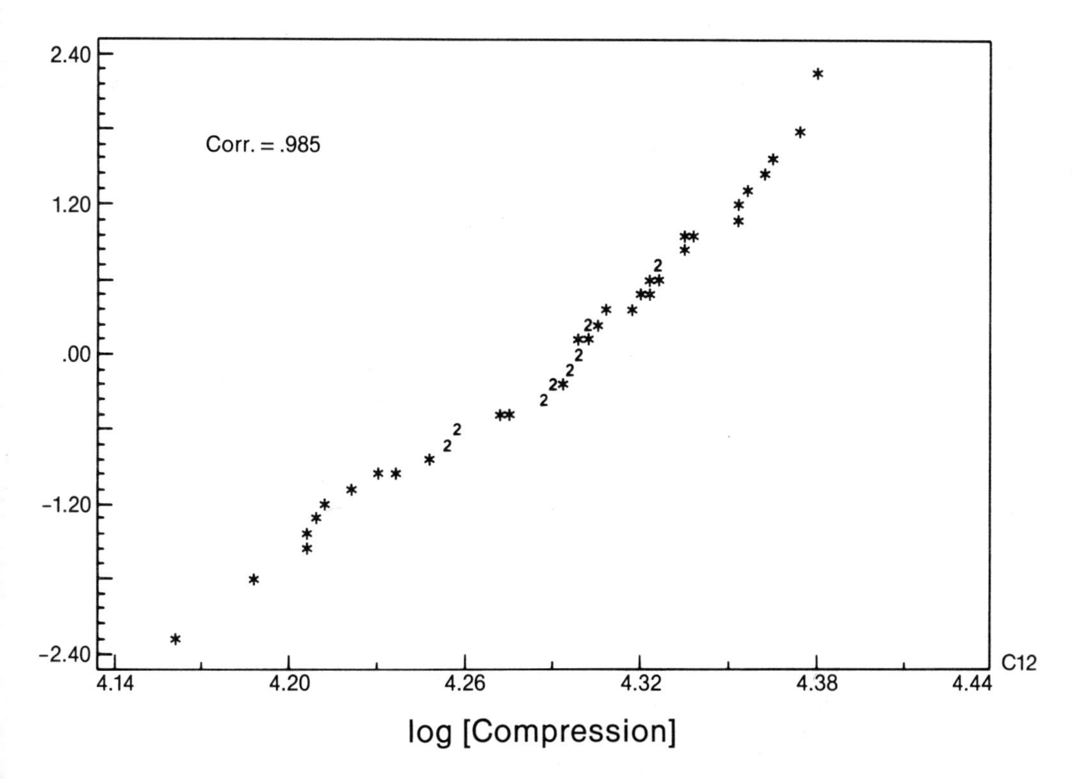


**Fig. 2(b). Lognormal Plot**

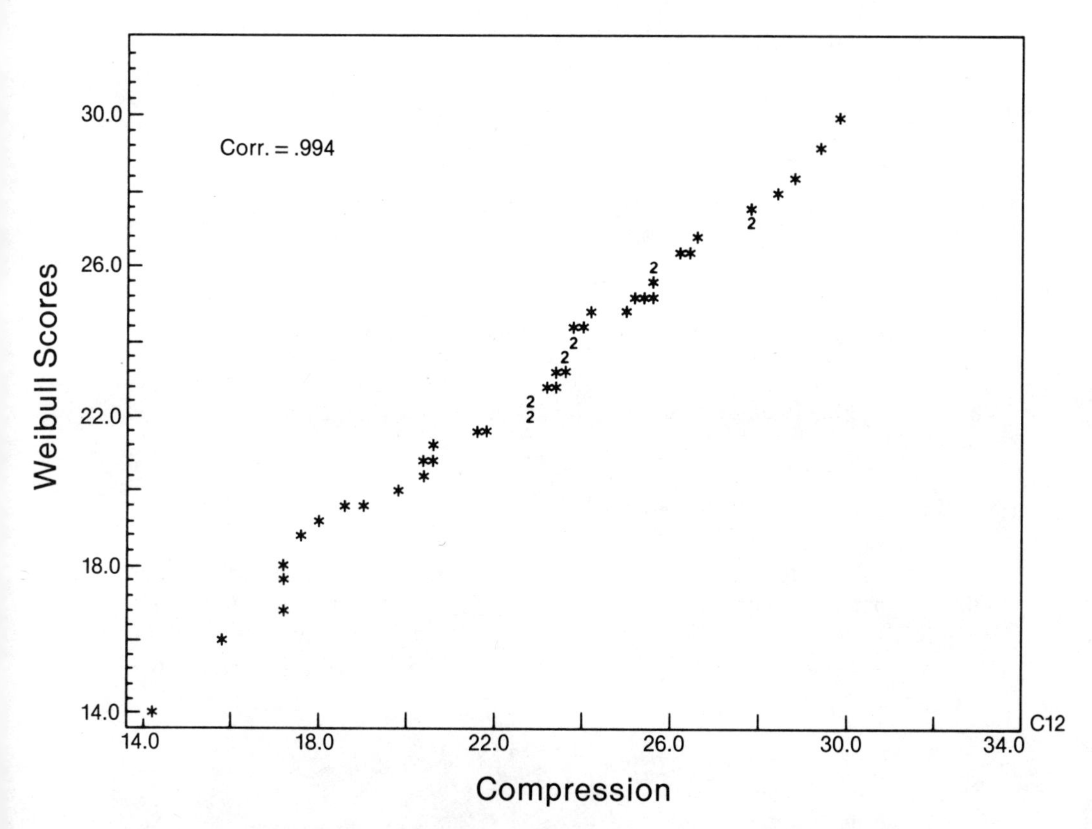


**Fig. 2(c). Weibull Plot**

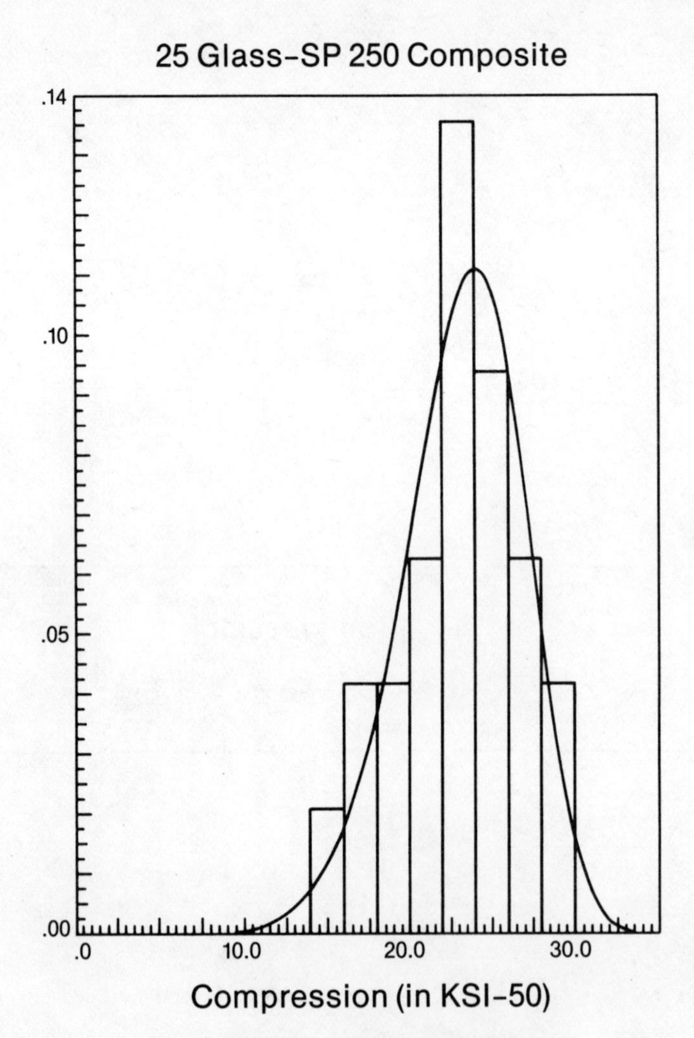


**Fig. 3. Weibull fit to the compression data.**

density on the histogram of Fig. 1. Figure 4 displays all three fitted distributions.

## Estimation of a Weibull Percentile

It is reasonable to set quality standards in terms of lower percentiles of the lifetime or strength distribution. The normal population theory, and hence lognormal is straight forward.

The three parameter Weibull distribution has percentiles defined by

$$\alpha = F(\xi_\alpha) = 1 - e^{-\left(\frac{\xi_\alpha - \beta}{\theta}\right)^p}, \; 0 < \alpha < 1$$

so its $100\alpha^{\text{th}}$ percentile is

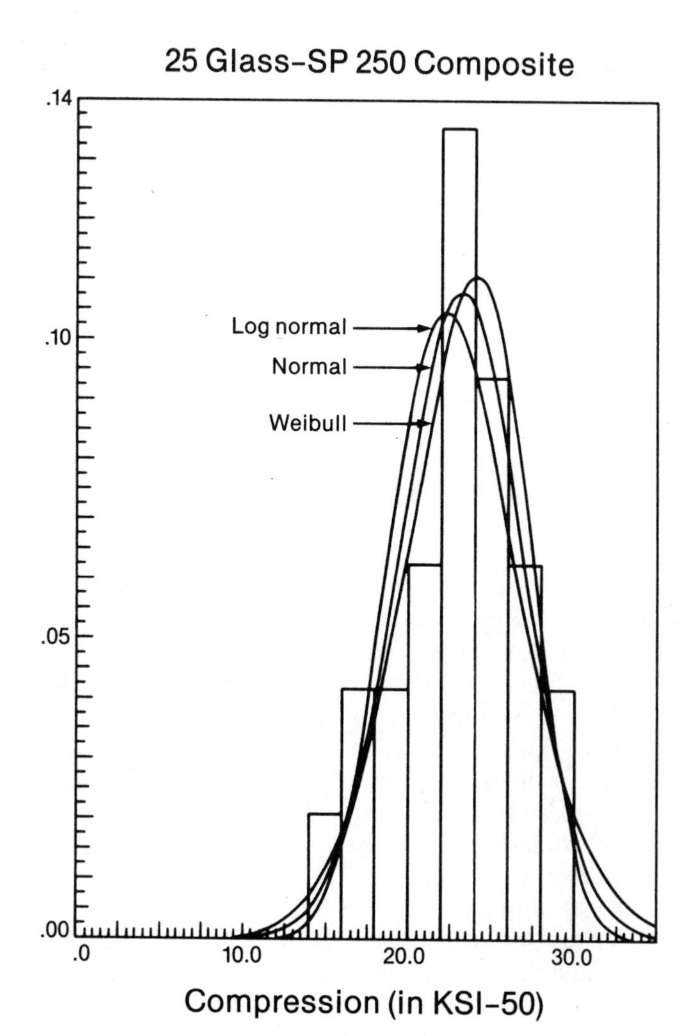


**Fig. 4. Normal, log-normal and Weibull fits to compression data.**

$$\xi_\alpha = [-ln(1-\alpha)]^{1/p}\theta + \beta.$$

The 10th percentile $\xi_{\cdot 10} = [-\ln(1-.1)]^{1/p}\,\theta + \beta$ has maximum likelihood estimator (m.l.e.)

$$\hat{\xi}_{\cdot 10} = [-\ln(.9)]^{1/\hat{p}}\,\hat{\theta} + \hat{\beta}$$

which is easy to compute once $\hat{p}$, $\hat{\theta}$ and $\hat{\beta}$ are obtained. Similar expressions hold for the m.l.e. of other percentiles. For the glass composite data $\hat{\xi}_{\cdot 10} =$ 68.267.

It is often preferable, for quality control purposes, to obtain a lower confidence bound for $\xi_\alpha$. If, for the 10th percentile, we find $\hat{\xi}_{\cdot 1,L}$ such that

$$P[\hat{\xi}_{\cdot 1,L} \leqslant \xi_{\cdot 1}] = .95$$

then, with probability .95, at least 90% of the population is above $\hat{\xi}_{\cdot 1,L}$.

As a large sample approximation, we take

$$\hat{\xi}_{\cdot 1,L} = \hat{\xi}_{\cdot 1} - 1.645 \text{ s}\hat{\text{d}}.\ (\hat{\xi}_{\cdot 1}) \tag{2.1}$$

where the estimated standard deviation, s$\hat{\text{d}}$. $(\hat{\xi}_{\cdot 1})$, depends on $\partial\xi_{\cdot 1}/\partial p$, $\partial\xi_{\cdot 1}/\partial\theta$, $\partial\xi_{\cdot 1}/\partial\beta$ and the asymptotic variances of $\hat{p}$, $\hat{\theta}$, and $\hat{\beta}$, all evaluated at the maximum likelihood values.

For the glass composite data (2.1) yields, $\hat{\xi}_{\cdot 1,L} = 56.894$ with 95% confidence.

Our studies indicate that quite large samples are required, say $n > 150$ or 200, in order to obtain accurate approximation to the nominal probabilities.

## 3. STRESS-STRENGTH MODELS WITH COVARIATES

To ascertain the reliability of an equipment or the viability of a material, one should bear in mind the stress conditions of the environment where the equipment or the material will be used. When the environmental stress $(X)$ is uncertain, it should be viewed as a random variable. The variability of the strength $(Y)$, being unavoidable in the real world, should also be modeled as a random quantity. In this stochastic formulation of stress versus strength, the prime target of study is the estimation of the reliability $R = P[Y > X]$. For some important nonparametric and parametric results concerning the estimation of reliability, in a stress-strength setting, the reader is referred to ([1]-[8]).

Often the experimenter has access to the measurements of some auxiliary variables which affect the strength of the material, and some others that influence the stress. Stimulated by practical situations which were brought to our attention by some scientists, we have embarked upon an important extension of the classical stress-strength model in order to explicitly incorporate the covariates into the model and develop appropriate analytical techniques.

*Example 1.* A board used in construction has strength $Y$ which can be obtained only by destructive testing. However, the stiffness $z$ is easily measured.

The data of strength measurements reported in Bhattacharyya and Johnson (1981) suggest that the strength $Y$ can be modeled as

$$Y = \alpha_2 + \beta_2 z + e_2$$

where $e_2$ is $N(0, \sigma_2^2)$. For a specimen of stiffness $z$

$$P[Y > X|z] = \Phi\left(\frac{\alpha_2 + \beta_2 z - \mu_1}{(\sigma_1^2 + \sigma_2^2)^{1/2}}\right).$$

where $\Phi$ denotes the standard normal cdf.

*Example 2.* Let $Y$ be the strength (in p.s.i.) of a glass fiber reinforced case, and let $X$ denote the operating pressure (stress). The data plots of Figs. (5a, b)

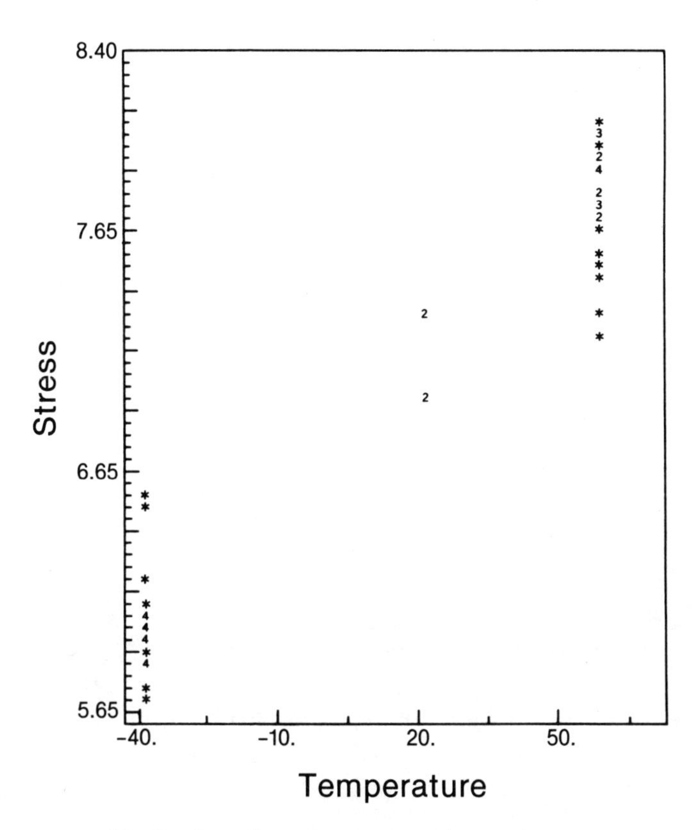


**Fig. 5a. Operating pressure versus temperature.**

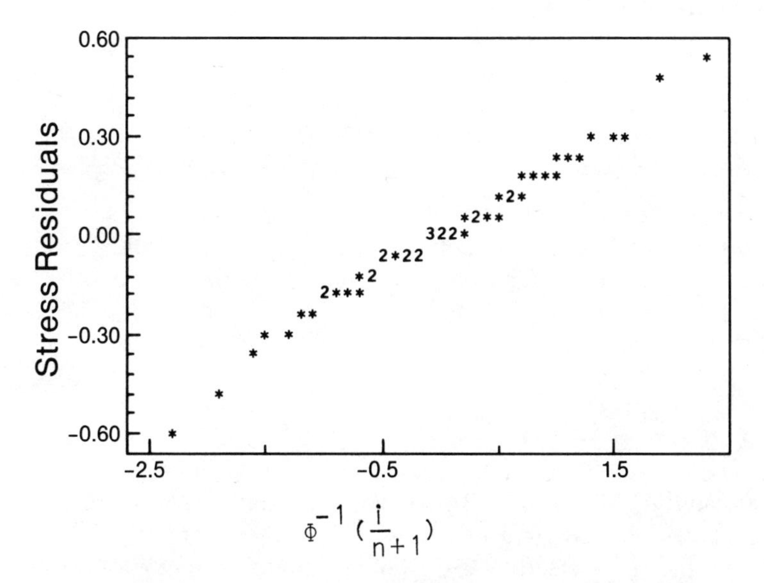


**Fig. 5b. Residuals versus their normal scores for the data of Fig. 5a.**

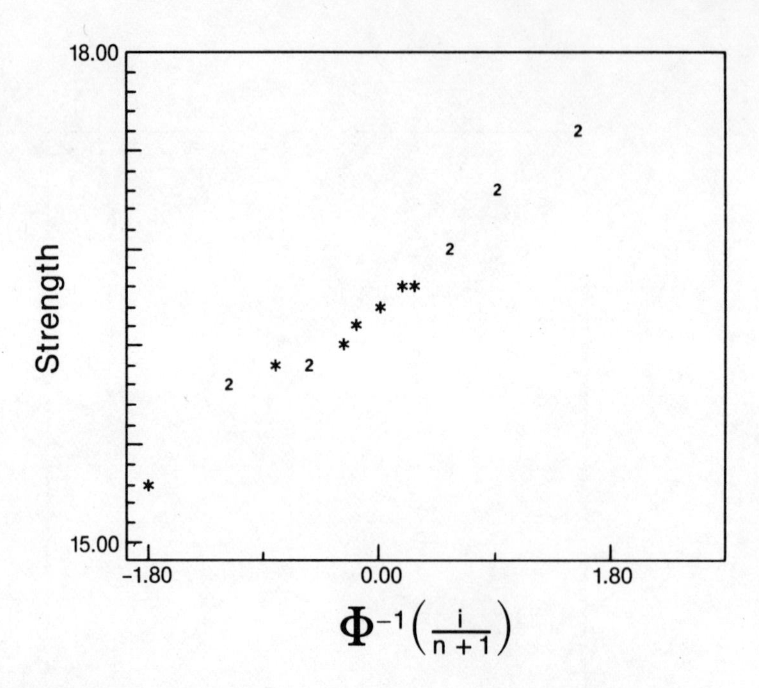


**Fig. 6. Strength (burst pressure) versus normal scores** $\Phi^{-1}\left(\frac{i}{n+1}\right)$. **n** = 17.

suggest that the stress depends on ambient temperature $z$ by a linear model

$$X = \alpha_1 + \beta_1 z + e_1$$

where $e_1$ is $N(0, \sigma_1^2)$. Also, the plot of Fig. 6 indicates that a normal model is reasonable for the strength.

For a given temperature $z$,

$$P[Y > X|z] = \Phi\left(\frac{\mu_2 - \alpha_1 - \beta_1 z}{(\sigma_1^2 + \sigma_2^2)^{1/2}}\right).$$

In both cases, the reliability has a similar structure.

Motivated by these examples we consider a general formulation that allows the stress $X$ to depend on $p_1$ number of covariates denoted by $\underset{\sim}{z}_1$, and strength $Y$ to depend on $p_2$ number of covariates denoted by $\underset{\sim}{z}_2$. We assume that a homoscedastic linear normal model holds, that is,

$$\begin{aligned} X|\underset{\sim}{z}_1 &= \underset{\sim}{\beta}_1' \underset{\sim}{z}_1 + e_1 \\ X|\underset{\sim}{z}_2 &= \underset{\sim}{\beta}_2' \underset{\sim}{z}_2 + e_2 \end{aligned} \tag{3.1}$$

where $\underset{\sim}{\beta}_1$ and $\underset{\sim}{\beta}_2$ are $p_1$ and $p_2$-component (column) vectors of regression parameters, respectively, and $e_1$ and $e_2$ are independent $N(0, \sigma^2)$ errors. The reliability at the specific setting $\underset{\sim}{z}_1$, $\underset{\sim}{z}_2$ of the covariates is then given by

$$R(\underset{\sim}{z}_1, \underset{\sim}{z}_2) = P[Y > X|\underset{\sim}{z}_1, \underset{\sim}{z}_2]$$

$$= P\left[\frac{(Y - \underset{\sim}{\beta}_2'\underset{\sim}{z}_2) - (X - \underset{\sim}{\beta}_1'\underset{\sim}{z}_1)}{\sqrt{2}\sigma} > \frac{\underset{\sim}{\beta}_1'\underset{\sim}{z}_1 - \underset{\sim}{\beta}_2'\underset{\sim}{z}_2}{\sqrt{2}\sigma}\right] \tag{3.2}$$

$$= \Phi(\delta)$$

where

$$\delta = (\underset{\sim}{\beta}_2'\underset{\sim}{z}_2 - \underset{\sim}{\beta}_1'\underset{\sim}{z}_1)/\sqrt{2}\sigma. \tag{3.3}$$

Based on the data $(X_i, \underset{\sim}{z}_{1i})$, $i = 1, \ldots, m$ and $(Y_j, \underset{\sim}{z}_{2j})$, $j = 1, \ldots, n$, we wish to make inferences on the reliability $R(\underset{\sim}{z}_1, \underset{\sim}{z}_2)$. Here we focus our attention on the Bayesian inference and present the main results which constitute an extension of a similar model treated in Enis and Geisser (1971). We introduce the following notations.

$$\begin{aligned}
\underset{\sim}{Z}_1' &= (\underset{\sim}{z}_{11}, \ldots, \underset{\sim}{z}_{1m}), \underset{\sim}{Z}_2' = (\underset{\sim}{z}_{21}, \ldots, \underset{\sim}{z}_{2n}) \\
\underset{\sim}{C}_1 &= \underset{\sim}{Z}_1'\underset{\sim}{Z}_1, \ \underset{\sim}{C}_2 = \underset{\sim}{Z}_2'\underset{\sim}{Z}_2 \\
\hat{\underset{\sim}{\beta}}_1 &= \underset{\sim}{C}_1^{-1}\underset{\sim}{Z}_1'\underset{\sim}{X}, \ \hat{\underset{\sim}{\beta}}_2 = \underset{\sim}{C}_2^{-1}\underset{\sim}{Z}_2'\underset{\sim}{Y} \\
N &= m + n, \ p = p_1 + p_2 \\
(N - p)s^2 &= (\underset{\sim}{X}'\underset{\sim}{X} - \hat{\underset{\sim}{\beta}}_1'\underset{\sim}{Z}_1'\underset{\sim}{X}) + (\underset{\sim}{Y}'\underset{\sim}{Y} - \hat{\underset{\sim}{\beta}}_2'\underset{\sim}{Z}_2'\underset{\sim}{Y}).
\end{aligned} \tag{3.4}$$

The sufficient statistics are $\hat{\underset{\sim}{\beta}}_1$, $\hat{\underset{\sim}{\beta}}_2$ and $s^2$, and the likelihood $L$ has the form

$$L \propto \sigma^{-N}\exp\left\{-\frac{1}{2\sigma^2}[(\hat{\underset{\sim}{\beta}}_1 - \underset{\sim}{\beta}_1)'\underset{\sim}{C}_1(\hat{\underset{\sim}{\beta}}_1 - \underset{\sim}{\beta}_1) + (\hat{\underset{\sim}{\beta}}_2 - \underset{\sim}{\beta}_2)'\underset{\sim}{C}_2(\hat{\underset{\sim}{\beta}}_2 - \underset{\sim}{\beta}_2) + (N - p)s^2]\right\}$$

With the vague prior $\frac{1}{\sigma}\, d\sigma d\underset{\sim}{\beta}_1\underset{\sim}{\beta}_2$, the posterior distribution is proportional to

$$\frac{1}{\sigma^{p_1+p_2}} e^{-\frac{1}{2\sigma^2}[(\hat{\underset{\sim}{\beta}}_1 - \underset{\sim}{\beta}_1)'\underset{\sim}{C}_1(\hat{\underset{\sim}{\beta}}_1 - \underset{\sim}{\beta}_1) + (\hat{\underset{\sim}{\beta}}_2 - \underset{\sim}{\beta}_2)'\underset{\sim}{C}_2(\hat{\underset{\sim}{\beta}}_2 - \underset{\sim}{\beta}_2)]}$$

$$\times \frac{1}{\sigma^{N-p-1}} e^{\frac{-(N-p)s^2}{2\sigma^2}}. \tag{3.5}$$

That is, conditional on $\sigma$ (and $\hat{\underset{\sim}{\beta}}_1$, $\hat{\underset{\sim}{\beta}}_2$, $s^2$), $\underset{\sim}{\beta}_1$ is distributed as a $p_1$-variate normal $N_{p_1}(\hat{\underset{\sim}{\beta}}_1, \sigma^2\underset{\sim}{C}_1^{-1})$ independent of $\underset{\sim}{\beta}_2$ which is distributed as $N_{p_2}(\hat{\underset{\sim}{\beta}}_2, \sigma^2\underset{\sim}{C}_2^{-1})$. Also

$$V \equiv \frac{(N - p)s^2}{\sigma^2} \text{ is } \chi^2_{N-p} \tag{3.6}$$

and, conditional on $\sigma$,

$$\delta = \frac{\beta_2' z_2 - \beta_1' z_1}{\sqrt{2}\sigma} \text{ is } N\left(\frac{\hat{\beta}_2' z_2 - \hat{\beta}_1' z_1}{\sqrt{2}\sigma}, \frac{1}{2}(z_1' C_1^{-1} z_1 + z_2' C_2^{-1} z_2)\right). \quad (3.7)$$

By (3.6) and (3.7), the posterior pdf of $(\delta, V)$ is the product of a normal pdf and a $\chi^2_{N-p}$ pdf.

Setting

$$c = \frac{1}{2}(z_1' C_1^{-1} z_1 + z_2' C_2^{-1} z_2)$$

$$k = \frac{\hat{\beta}_2' z_2 - \hat{\beta}_1' z_1}{\sqrt{2}\,\nu s}, \quad \nu = N - p$$

the posterior pdf of $\delta$ becomes

$$p_\Delta(\delta) = \frac{1}{\sqrt{2\pi c}} \frac{\left(\frac{k^2}{c} + 1\right)^{-\nu/2}}{\Gamma\left(\frac{\nu}{2}\right)} e^{-\delta^2/2c} \sum_{j=0}^{\infty} \left[\frac{\sqrt{2}}{c} k\delta \left(1 + \frac{k^2}{c}\right)^{-1/2}\right]^j \frac{\Gamma\left(\frac{\nu+j}{2}\right)}{\Gamma(j+1)}. \quad (3.8)$$

The posterior mean and variance are

$$E_{post.}(\delta) = \sqrt{2}\, k\, \Gamma\left(\frac{\nu+1}{2}\right) \Big/ \Gamma\left(\frac{\nu}{2}\right)$$

$$\mathrm{Var}_{post.}(\delta) = c + k^2 \left[\nu - \frac{2\Gamma^2\left(\frac{\nu+1}{2}\right)}{\Gamma^2\left(\frac{\nu}{2}\right)}\right]. \quad (3.9)$$

The posterior pdf of the reliability $R(z_1, z_2) = \Phi(\delta)$ can then be expressed as

$$p_R(r) = p_\Delta(\Phi^{-1}(r)) \frac{1}{\phi(\Phi^{-1}(r))} \quad (3.10)$$

where $\phi$ denotes the $N(0, 1)$ pdf.

In the preceding general development we have allowed a vector of covariates with each of the stress and strength random variables. When only one of them has covariates and the other has none, appropriate modifications of our results can readily be made.

Referring to the data of Example 2, where temperature is the only covariate associated with the stress, we see that the stress distribution and strength distribution are widely separated giving extremely high reliabilities. For illustrative purposes, we take $\bar{y} = 8.4853$ and pool discrepant estimates of variance to obtain $s = .3457$. The other quantities are $m = 51$, $n = 17$, $\hat{\alpha} = .2152$ and $\hat{\beta} =$

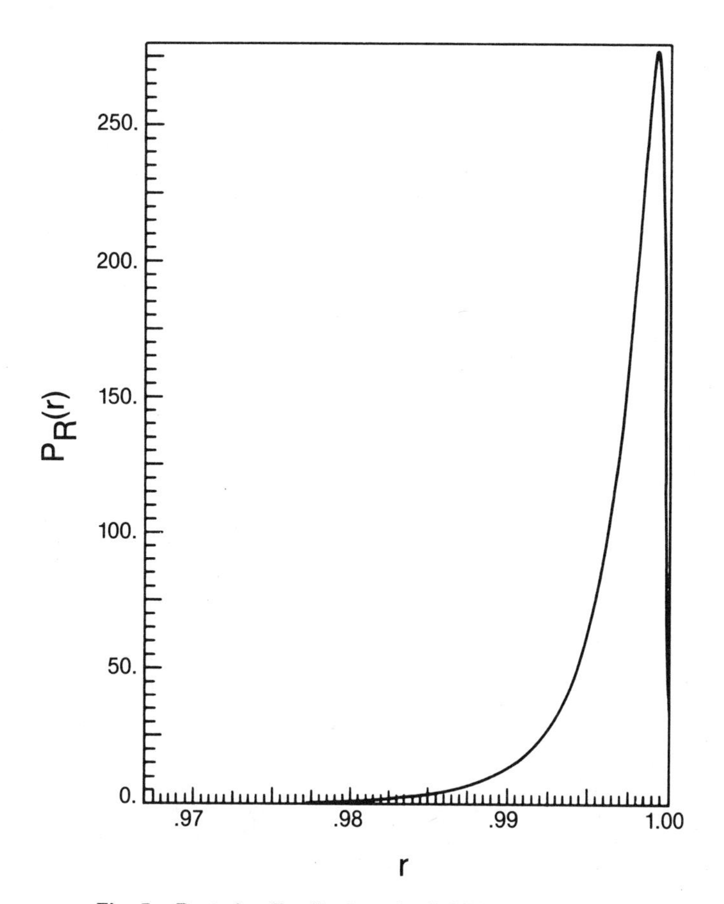


**Fig. 7a. Posterior distribution of reliability at $z = 21°C$.**

.0182. The posterior distributions of the reliability $R$ corresponding to the temperatures $z = 21°C$ and $59°C$ are obtained on a computer and are presented in Figs. 7a and 7b.

Also of interest is the problem of obtaining simultaneous lower confidence bounds for the reliability, that is, the bounds that hold for an entire range of values of the covariates with a specified confidence probability. For simplicity, we confine ourselves to the situation of Example 2 where $X$ has the single covariate $z$, and $Y$ has none. Here again, we assume that $\sigma_1 = \sigma_2$ and denote their common value by $\sigma$. Also, we write $\mu$, $\alpha$ and $\beta$ for the parameters $\mu_2$, $\alpha_1$ and $\beta_1$ of Example 2.

Let us denote

$$c_1 = (m^{-1} + n^{-1})^{1/2}, \; c_2 = \left[\sum_{i=1}^{m}(z_i - \bar{z})^2\right]^{-1/2}. \tag{3.11}$$

Then, conditionally given $\sigma$, the posterior distribution of $\zeta_1 \equiv (\mu - \alpha)/(\sigma c_1)$ and $\zeta_2 \equiv \beta/(\sigma c_2)$ is normal with means $(\bar{y} - \bar{x})/(\sigma c_1)$, and $\hat{\beta}/(\sigma c_2)$, unit variances, and correlation zero. Further, $V \equiv (N - 3)s^2/\sigma^2$ is distributed as $\chi^2_{N-3}$.

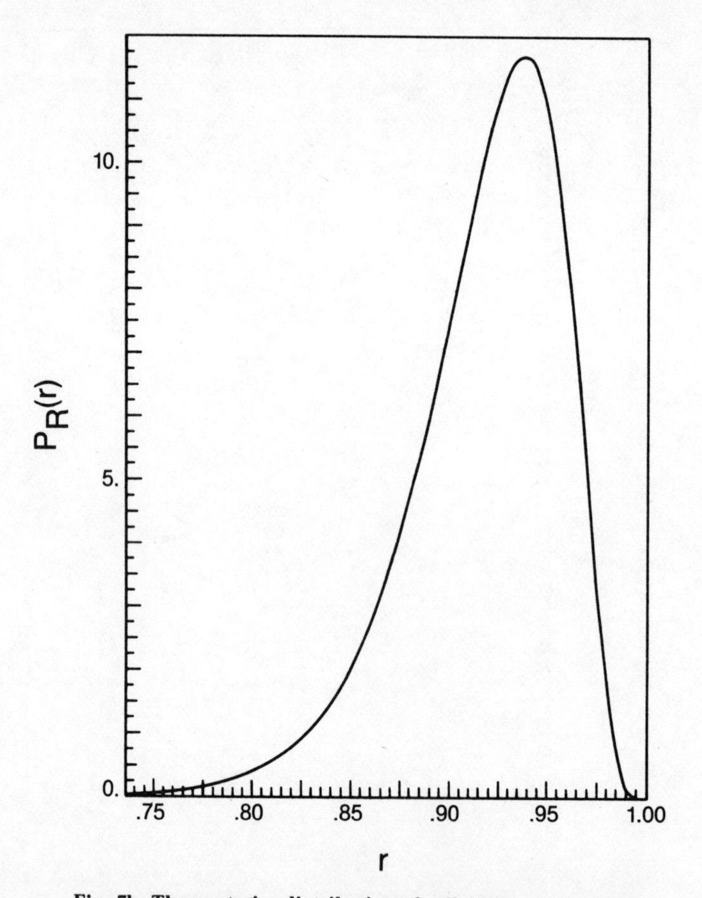


**Fig. 7b. The posterior distribution of reliability for z = 59°C.**

Considering the linear combination

$$c_1\zeta_1 + c_2(z - \bar{z})\zeta_2 = [\mu - \alpha - \beta(z - \bar{z})]/\sigma,$$

an application of the Cauchy-Schwarz inequality gives

$$\left(\frac{\mu - \alpha - \beta(z - \bar{z})}{\sigma} - \frac{\bar{y} - \bar{x} - \hat{\beta}(z - \bar{z})}{s}\right)^2 \leqslant [c_1^2 + c_2^2(z - \bar{z})^2]\,U \quad (3.12)$$

for all $z$, where

$$U = \left(\frac{\mu - \alpha}{\sigma c_1} + \frac{\bar{y} - \bar{x}}{sc_1}\right)^2 + \left(\frac{\beta}{\sigma c_2} - \frac{\hat{\beta}}{sc_2}\right)^2. \quad (3.13)$$

Now, conditionally given $\sigma$, $U$ has the noncentral $\chi^2$ distribution with 2 degrees of freedom, and the noncentrality parameter

$$\eta = \left(\frac{V}{N-3} - 1\right)^2 s^{-2}\left[\frac{(\bar{y} - \bar{x})^2}{c_1^2} + \frac{\hat{\beta}^2}{c_2^2}\right]. \quad (3.14)$$

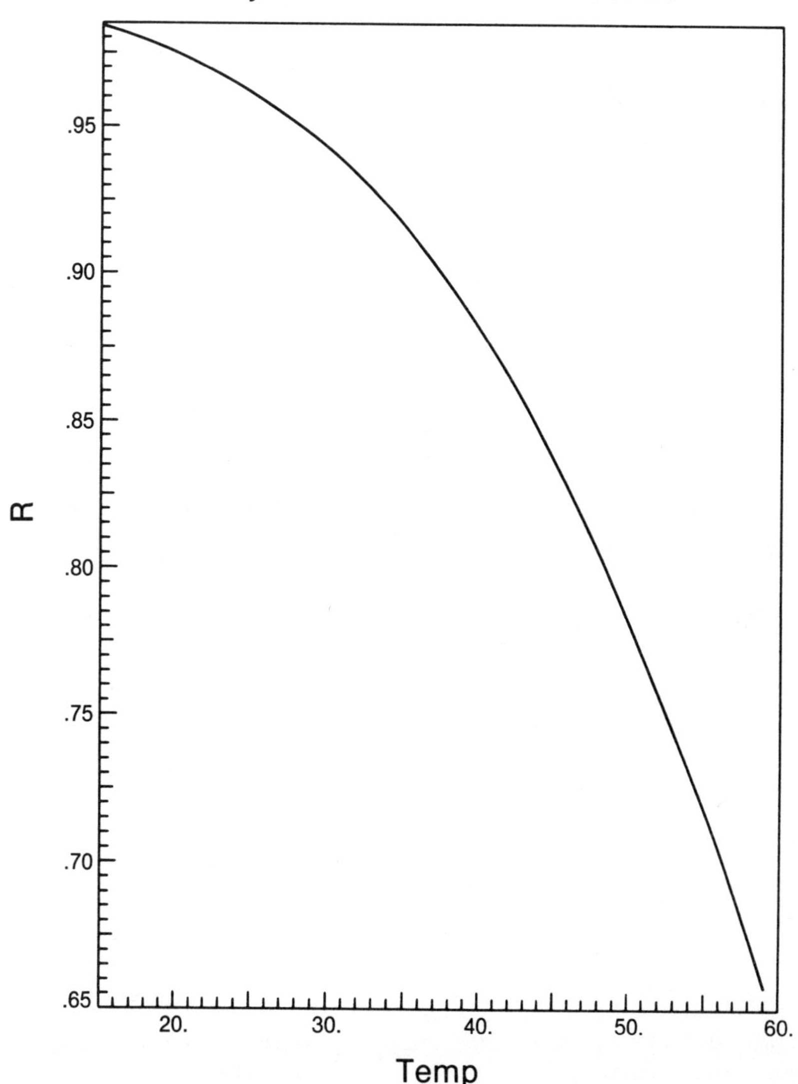


**Fig. 8. Simultaneous 95% lower confidence bounds.**

Noting that $\eta$ is a function of $V$ which is distributed as central $\chi^2_{N-3}$, we have

$$P[U \leqslant u] = \int_0^\infty P[U \leqslant u|\sigma] f_V(\mathrm{v})\,d\mathrm{v}$$

$$= \sum_{j=0}^{\infty} \int_0^u f_{2+2j}(y)\,dy \left\{ \int_0^\infty e^{-\eta/2} \frac{(\eta/2)^j}{j!} f_{N-3}(\mathrm{v})\,d\mathrm{v} \right\} \tag{3.15}$$

where $f_\nu(\cdot)$ denotes the pdf of central $\chi^2_\nu$ distribution. Note that $\eta$ appearing in the integrand of the last expression is actually a function of v, as shown in (3.14).

In order to obtain 95% simultaneous lower confidence bounds for the reliability $R(z) = P[Y > X|z]$, we define $u_{.05}$ by $P[U \leqslant u_{.05}] = .95$. This can be determined by numerical solution on a computer. Referring to (3.12) we then have

$$\frac{\mu - \alpha - \beta(z - \bar{z})}{\sigma} \geqslant \frac{\bar{y} - \bar{x} - \hat{\beta}(z - \bar{z})}{s} - u_{.05}^{1/2} \left[c_1^2 + c_2^2 (z - \bar{z})^2\right]^{1/2} \quad (3.16)$$

simultaneously for all $z$, with 95% posterior probability. Consequently, the 95% simultaneous lower confidence bounds for $R(z)$ are given by

$$R(z) \geqslant \Phi(q(z))$$

where

$$q(z) = 2^{-1/2}\left\{\frac{\bar{y} - \bar{x} - \hat{\beta}(z - \bar{z})}{s} - u_{.05}^{1/2}\left[\frac{1}{m} + \frac{1}{n} + \frac{(z - \bar{z})^2}{\Sigma(z_i - \bar{z})^2}\right]^{1/2}\right\}.$$

For the data of Example 2, a computation with double precision arithmetic yields $u_{.05} = 21.6$. The curve of Bayesian simultaneous 95% lower confidence bounds for $R(z)$ is shown in Fig. 8.

## REFERENCES

G.K. Bhattacharyya and R.A. Johnson (1974). Estimation of reliability in a multicomponent stress-strength model. *J. Amer. Statist.* Assoc., **69**, 966-70.

G.K. Bhattacharyya and R.A. Johnson (1975). Stress-strength models for system reliability. *Proc. Symp. on Reliability and Fault-tree Analysis.* SIAM, 509-32.

G.K. Bhattacharyya and R.A. Johnson (1977). Estimation of system reliability by nonparametric techniques. *Bulletin of the Mathematical Society of Greece* (memorial Volume), 94-105.

G.K. Bhattacharyya and R.A. Johnson (1981). Stress-strength models for reliability: overview and recent advances. *Proc. 26th Design of Experiments Conference,* 531-548.

Z.W. Birnbaum and R.C. McCarty (1958). A distribution free upper confidence bound for $P(Y < X)$ based on independent samples of $X$ and $Y$. *Ann. Math Statist.*, **29**, 558-62.

P. Enis and S. Geisser (1971). Estimation of the probability that $Y < X$. *J. Amer. Statist. Assoc.*, **66**, 162-8.

Z. Govindarajulu (1968). Distribution-free confidence bounds for $P(X < Y)$. *Ann. Inst. Statist. Math.,*, **20**, 229-38.

D.B. Owen, K.J. Craswell, and D.L. Hanson (1964). Nonparametric upper confidence bounds for $P(Y < X)$ and confidence limits for $P(Y < X)$ when $X$ and $Y$ are normal. *J. Amer. Statist. Assoc.,*, **59**, 906-24.

# A Decreasing Failure Rate, Mixed Exponential Model Applied to Reliability*

Janet M. Myhre

*Claremont McKenna College*

**Abstract**

Decreasing failure rates for electronic equipment used on the Polaris, Poseidon and Trident missile systems have been observed. The mixed exponential distribution has been shown to fit the life data for the electronic equipment on these systems. This paper discusses some of the estimation problems which occur with the decreasing failure rate mixed exponential distribution when the test data is censored and only a few failures are observed. For these cases sufficient conditions are obtained that maximum likelihood estimators of the shape and scale parameters for the distribution exist. Actual data, obtained from the testing of missile electronic packages, are provided to illustrate these concepts and verify the applicability and usefulness of the techniques described.

## 1. INTRODUCTION

There have been only a few parametric models extensively examined for application to reliability; these include the exponential distribution of Epstein-Sobel [6], the Weibull distribution [16], and the fatigue model of Birnbaum-Saunders [4]. The one most widely utilized for electronic components has been the exponential model, not only because of its simple and intuitive properties but also because of the extent of the estimation and sampling procedures which have been developed from the theory.

*Research, in part, supported by the Office of Naval Research Contract N00014-76-C-0695.

One of the early discoveries was that mixtures of exponentially distributed random variables have a decreasing failure rate, see [13]. Thus any two groups of components with constant, but different, failure rates would, if mixed and sampled at random, exhibit a decreasing failure rate. As a consequence, the family of life lengths with decreasing failure rate certainly arises in practice and particular subsets of this family could be of great utility for specific applications, see e.g., Cozzolino [5]. We examine one such model with shape and scale parameters, call them $\alpha$ and $\beta$ respectively, which is based upon a particular mixture of exponential distributions. This family was introduced by Afanas'ev [1] and later by Lomax [10] as a generalization of a Pareto distribution. Section 3 compares this mixed exponential distribution to the exponential distribution using data from Poseidon flight control packages.

Kulldorff and Vännman [9] and Vännman [15] have studied a variant of this mixed exponential model containing a location parameter. They obtained a best linear unbiased estimate of the scale parameter assuming that the shape parameter, call it $\alpha$, was known and in a region restricted so that both the mean and the variance exist, namely $\alpha > 2$. When this restriction of $\alpha > 2$ cannot be met an estimate based on a few order statistics, which are optimally spaced, is claimed to be an asymptotically best linear unbiased estimate and tables of the weights as functions of the number of spacings are provided. In all cases, the shape parameter was assumed known and the sample was either complete or type II censored. It is contended that BLUE estimates of the shape parameter are not attainable.

Harris and Singpurwalla [7] examined the method of moments as an estimation procedure for this same model but again with the shape parameter restricted to $\alpha > 2$ and with a complete sample.

In both papers [9] and [7], it is stated that maximum likelihood estimates are difficult to obtain. In a later paper Harris and Singpurwalla [8] exhibit the maximum likelihood equations for complete samples.

In this paper the maximum likelihood estimates for both the shape and scale parameters are obtained, jointly and separately, with simple sufficient conditions given for their existence. These estimates are derived for censored data (and *a fortori* for complete samples) even with a paucity of failure observations, namely one.

The existence conditions obtained here for the maximum likelihood estimates apply even to the case where the variance and possibly the mean do not exist: $0 < \alpha < 2$. Moreover, the estimates of the shape parameter $\alpha$ which have been obtained from actual data indicate that this region $0 < \alpha < 2$ is important because all the estimates obtained of $\alpha$ have been less than unity.

## 2. MODEL

We postulate that the underlying process which determines the length of life of the component under consideration is the following: The quality of construction determines a level of resistance to stress which the component can tolerate. The service environment provides shocks of varying magnitude to the component, and failure takes place when, for the first time, the stress from an environmentally induced shock exceeds the strength of the component.

If the time between shocks of any magnitude is exponentially distributed with a mean depending upon that magnitude then the life length of each component will be exponentially distributed with a failure rate which is determined by the quality of assembly. It follows that each component has a constant failure rate but that the variability in manufacture and inspection techniques forces some components to be extremely good while a few others are bad and most are in between.

Let $X_\lambda$ be the life length of a component in such a service environment, with a constant failure rate $\lambda$ which is unknown. The variability of manufacture determines various percentages of the $\lambda$-values and this variability can be described by some distribution, say G.

Let $T$ be the life length of one of the components which is selected at random from the population of manufactured components. We denote the reliability of this component by $R$ and we have

$$R(t) = P[T > t] \quad \text{for } t > 0.$$

Let $\Lambda$ be the random variable which has distribution $G$. We can write

$$R(t) = E_\Lambda P[X_\lambda > t | \Lambda = \lambda] = \int_0^\infty e^{-\lambda t} dG(\lambda). \tag{1}$$

Because of having a form which can fit a wide variety of practical situations when both scale and shape parameters are disposable, it is assumed that $G$ is a gamma distribution, i.e., for some $\alpha > 0$, $\beta > 0$,

$$g(\lambda) = \frac{\lambda^{\alpha-1} e^{-\lambda/\beta}}{\Gamma(\alpha)\, \beta^\alpha} \quad \text{for } \lambda > 0.$$

That this assumption is robust, even when mixing as few as five equally weighted $\lambda$'s, has been shown by recent work of Sunjata in an unpublished thesis [14]. It follows from Eq. (1) that the reliability function is

$$R(t) = \frac{1}{(1 + t\beta)^\alpha} = e^{-\alpha \ln(1 + t\beta)}. \tag{2}$$

The failure rate, hazard rate, can be shown to be

$$q(t) = \frac{\alpha\beta}{1 + t\beta}, \tag{3}$$

which is a decreasing function of $t > 0$.

Maximum likelihood estimates for $\alpha$, $\beta$ and hence $R(t)$ and $q(t)$ are given in Section 5.

## 3. A COMPARISON OF THE MIXED EXPONENTIAL WITH EXPONENTIAL USING REAL DATA

Data has been accumulating for years in the assessment of the reliability of electronic equipment for which there was no adequate statistical model. The following difficulties were recognized by practitioners: 1. The assumption of constant or increasing failure rate seemed to be incorrect. 2. However, the design of this electronic equipment indicated that individual items should exhibit a constant failure rate. A mixed exponential life distribution accounts for both the design knowledge and the observed life lengths. Maximum likelihood procedures allow for joint estimation of the parameters of this distribution in the most commonly encountered situation where complete data is not available.

We now give some actual data sets from two different lots of Poseidon flight control electronic packages which illustrate these points. Each package has recorded, in minutes, either a failure time or an alive time. An alive time is sometimes called a "run-out" and is the time the life test was terminated with the package still functioning.

First Data Set
Failure times: 1, 8, 10
Alive times: 59, 72, 76, 113, 117, 124, 145, 149, 153, 182, 320.

Second Data Set
Failure times: 37, 53
Alive times: 60, 64, 66, 70, 72, 96, 123.

If we assume that the data are observations from an exponential distribution (constant failure rate $\lambda$) then using the total life statistic, we have the estimates of reliability given in the left hand side of the table. If we assume that the data are observations from the mixed exponential distribution of equation (2) then using estimation techniques derived subsequently in this paper we have the estimates for reliability given in the right hand side of the table. Looking at the data from the two sets we would expect that at least for the first fifty minutes the reliability estimate for the second set of data would be *higher* than the reliability estimate for the first set of data, because in the first set 3 failures out of 14 trials have occured in the first ten minutes while in the second set only 1 failure out of 9 trials has occurred in the first fifty minutes. However, under the exponential assumption the reliability estimates for the first data set are consistently higher. Note that the mixed exponential estimates

**Table 1**

| | Exponential estimate of reliability | | Mixed exponential estimate of reliability | |
|---|---|---|---|---|
| time $t$ in min | Set 1 $\hat{R}_1(t)$ | Set 2 $\hat{R}_2(t)$ | Set 1 $\hat{R}_1(t)$ | Set 2 $\hat{R}_2(t)$ |
| 6 | .988 | .981 | .915 | .976 |
| 10 | .980 | .969 | .896 | .961 |
| 30 | .943 | .911 | .855 | .896 |
| 50 | .906 | .856 | .836 | .843 |
| 100 | .821 | — | .810 | — |
| 130 | .774 | — | .801 | — |
| $\lambda$: | .00017 | .00312 | $\hat{\alpha}$: .0453 | .420 |
| | | | $\hat{\beta}$: 1.03 | .01 |

are more consistent with what the data show; that is, for at least the first 50 minutes we expect the reliability estimate for the second set of data to be higher than the reliability estimate for the first set of data. Beyond this time, however, say at 100 minutes, the data indicate that the reliability estimate from the first set of data should be higher than the reliability estimate from the second set of data. Using mixed exponential estimates this is the case.

A statistical test to determine whether the data require a constant or decreasing failure rate was run on the data from Sets 1 and 2. For data Set 1 we reject constant failure rate in favor of decreasing failure rate at the .10 level. For data Set 2 we cannot reject the constant failure rate assumption. In this case, however, the constant failure rate estimates for reliability and the mixed exponential estimates for reliability are close. For data Set 2 one should not estimate reliability much beyond about 70 minutes since we do not have data to support those estimates.

## 4. RESIDUAL LIFE PROPERTY OF THE MODEL

An important property of this model is that residual life on a component is distributed as a mixed exponential. Thus a "burn-in" test of a component will yield a residual life which is also in the same family. This property seems to be shared only with the exponential among common parametric families of life distributions.

The residual life $T_h$ of a component is defined to be the life remaining after time $h$, given that the component is alive at time $h$. It can be shown that:

A burn-in for $h$ units of time on a component with initial life determined by a mixed exponential distribution with parameters $\alpha$ and $\beta$ will yield a residual life $T_h$ and will be distributed as a mixed exponential with parameters $\alpha$ and $\frac{\beta}{1+\beta h}$.

If follows that this life length model is "used better than new" or "new worse than used" in the sense that we have stochastic inequality between a new component and one that has been burned in, namely

$$T \overset{st}{\leqslant} T_h \text{ for all } h < 0.$$

An important consequence of this property is that one can calculate the value of the increased reliability attained by burn-in procedures as compared with the cost of conducting them. It has long been the practice to burn in electronic components based on intuitive ideas of "infant mortality" in order to provide reasonable assurance of having detected all defectively assembled units. This model, whenever it is applicable, makes possible an economic analysis. A variation of this result has been discussed in [3].

### Example

As an example of the applicability of this property, consider test data from Trident flight control packages.

Assume that burn-in data is distributed as a mixed exponential with shape parameter $\alpha$ and scale parameter $\beta$. These parameters were estimated (formulas in Section 5) to be

$$\hat{\alpha} = .57$$
$$\hat{\beta} = .0104$$

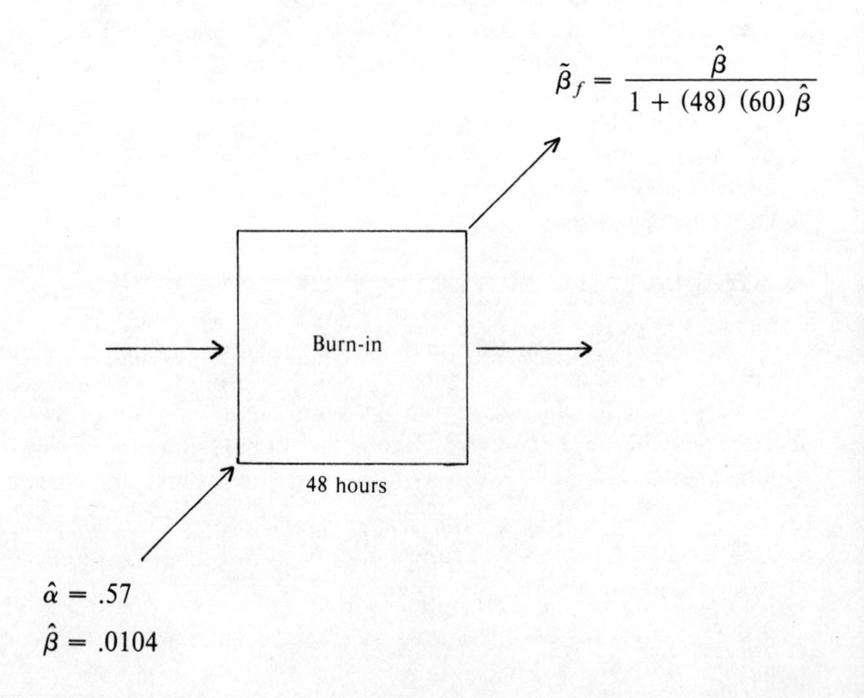


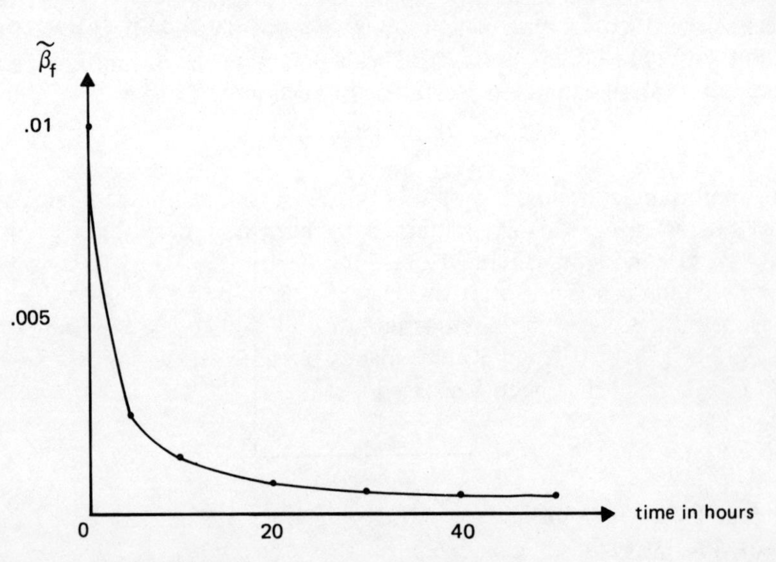


Forecast $\tilde{\beta}_f$ as a function of burn-in time

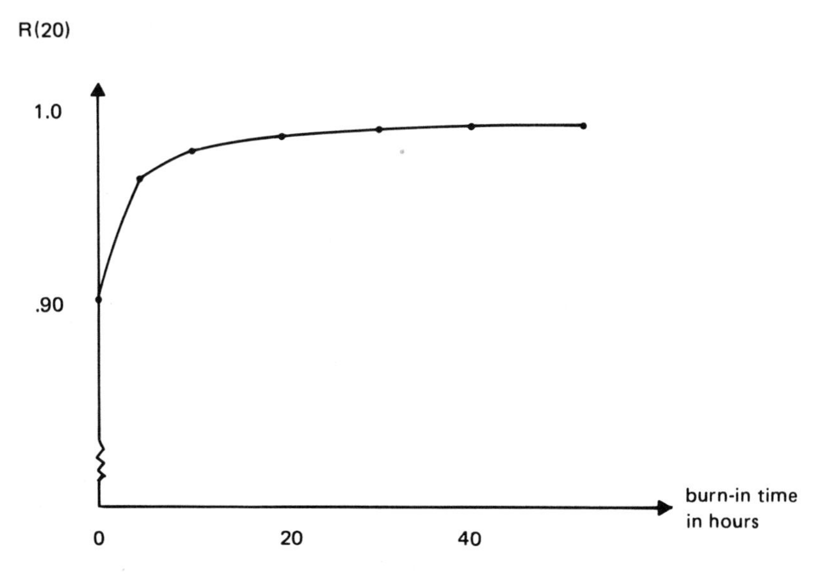


Estimated Burn-in Reliability at 20 minutes as a function of previous burn-in time

After 48 hours of burn-in the residual life $T_{48\text{ hours}}$ should be mixed exponential with parameters $\alpha$ and $\dfrac{\beta}{1 + 2880\beta}$. The first graph shows the change in $\tilde{\beta}_f$ as a function of burn-in hours. The second graph shows the change in estimated reliabilities at 20 minutes as a function of burn-in time, where reliability at time 20 minutes is estimated to be

$$\hat{R}(t) = [1 + 20\hat{\beta}_f]^{-\hat{\alpha}}.$$

Data consistent with success and failure data was obtained from another test called Pre-Test. We assume that the time to failure $T$ of flight control packages subjected to this type of test environment follows a mixed exponential distribution with shape parameter $\alpha$ and scale parameter $\beta$. Using Pre-Test data these parameters were estimated (formulas in Section 5) to be

$$\hat{\alpha} = .5739$$

$$\hat{\beta} = .1106$$

Pre-Test

60 minutes

$\hat{\alpha} = .5739$

$\hat{\beta} = .1106$

After 60 minutes of test the residual life $T_{60}$ should be a mixed exponential with parameters $\alpha$ and $\frac{\beta}{1+60\beta}$. We estimate these parameters by

$$\hat{\alpha} = .5739$$

$$\tilde{\beta}_f = \frac{\hat{\beta}}{1+t\hat{\beta}} = \frac{.1106}{1+60(.1106)} = .0145$$

$$\tilde{\beta}_f = \frac{\hat{\beta}}{1+60\hat{\beta}} = .0145$$

$$\hat{R}_f(1) = .99$$

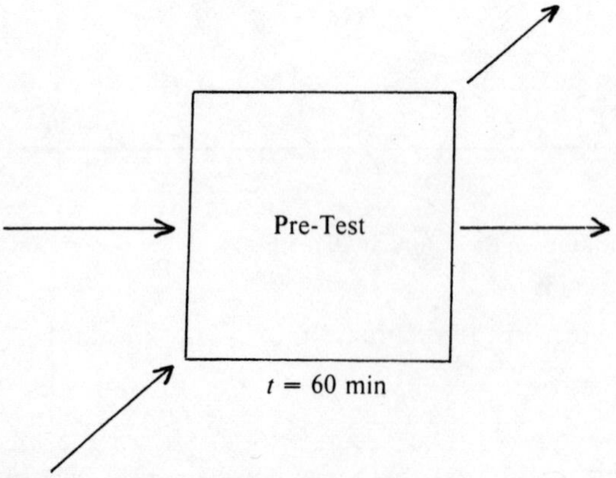

Pre-Test

$t = 60$ min

$$\hat{\alpha} = .5739$$

$$\hat{\beta} = .1106$$

$$\hat{R}(1) = .94.$$

Since $R(t) = (1+\beta t)^{-\alpha}$, we estimate reliability at 1 minute for a package which has not gone through Pre-Test to be

$$\hat{R}_{\hat{\beta}}(60) = [1 + (.1106)\ (1)]^{-.5739} = .94$$

We estimate reliability at 1 minute for a package which has gone through Pre-Test to be

$$\hat{R}_{\hat{\beta}_f}(60) = [1 + (.0145)\ (1)]^{-.5739} = .99.$$

Note that these reliabilities are for Pre-Test environments.

Now consider the entire screen test scheme for Trident flight control packages.

Note: All Job Stack tests are the same.

Effects of various environmental or burn-in tests (in the sense of reliability gain) can be estimated by comparing reliability estimates forecast at the end

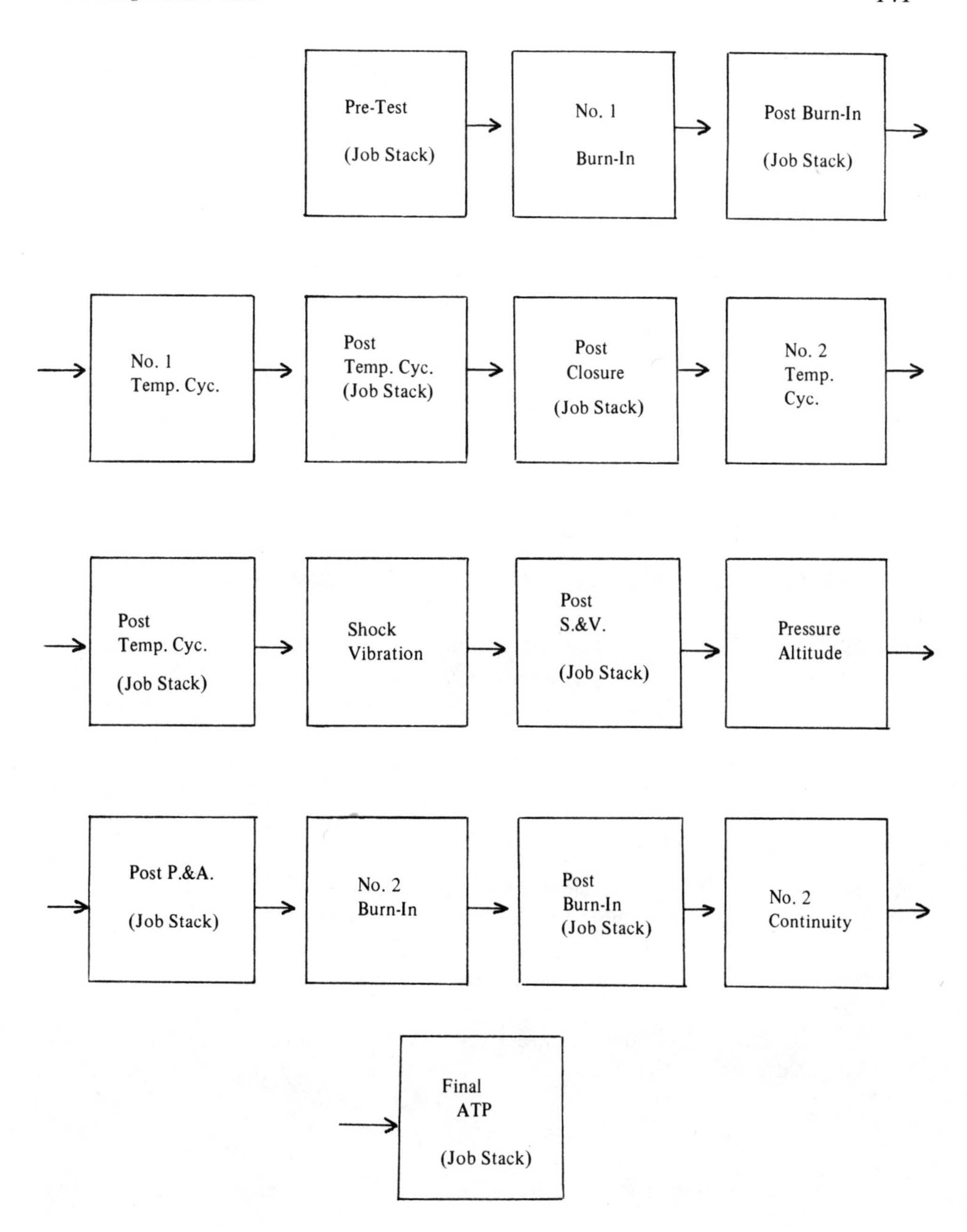


of the Job Stack preceding the environment to reliability estimates for the Job Stack following the environment.

For example

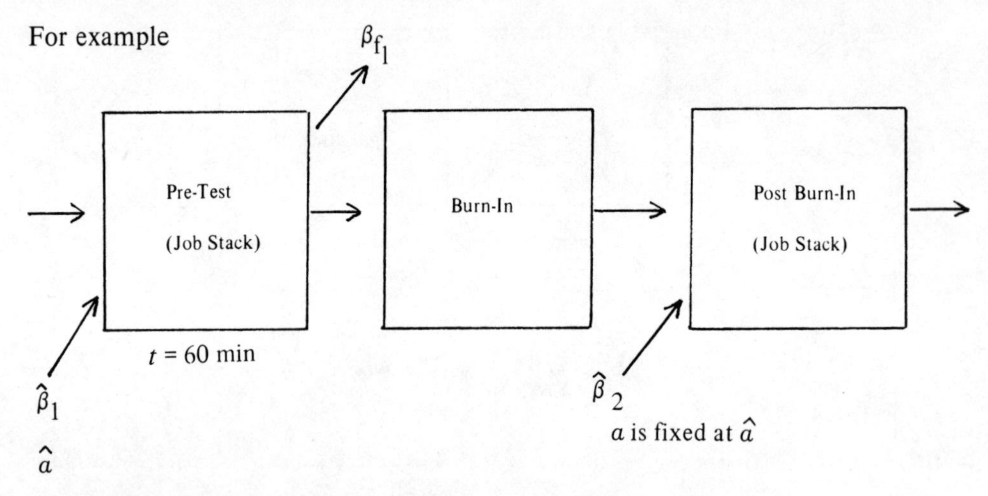


If burn-in is effective then we would expect that $\hat{\beta}_2 < \tilde{\beta}_{f_1}$.

## 5. ESTIMATION OF PARAMETERS WITH CENSORED DATA

Let us assume throughout this section that we are given $t_1, \ldots, t_k$ as observed times of failure while $t_{k+1}, \ldots, t_n$ are observed alive-times both obtained from a mixed exponential $(\alpha,\beta)$ life distribution with $1 \leqslant k \leqslant n$. We define two functions for $x > 0$.

$$S_1(x) = \sum_{i=1}^{n} \ln(1 + t_i x), \; S_2(x) = \frac{1}{k} \sum_{i=1}^{k} (1 + t_i x)^{-1}.$$

A result on the maximum likelihood estimation (m.l.e.) of the unknown parameters is now given which utilizes data of this type.

Theorem: Under the assumptions and conditions given

(i) When $\beta > 0$ is known, there exists a unique m.l.e. of $\alpha$, say $\hat{\alpha}$, given explicitly by

$$\hat{\alpha} = k/S_1(\beta).$$

(ii) When $\alpha > 0$ is known, there exists a unique m.l.e. of $\beta$, say $\hat{\beta}$, given explicitly by

$$\hat{\beta} = A^{-1}(0)$$

where $A$ is the monotone decreasing function defined by

$$A(x) = kS_2(x) - \alpha x S_1'(x) \text{ for } x > 0$$

with primes denoting derivatives.

(iii) When $\alpha,\beta$ are both unknown, the m.l.e. of $\beta$, say $\hat{\beta}$, is given implicitly, when it exists positively and finitely, by

$$\hat{\beta} = B^{-1}(0)$$

where $B$ is the function defined by

$$B(x) = \frac{S_2(x)}{x} - \frac{S'_1(x)}{S_1(x)} \text{ for } x > 0$$

and the m.l.e. of $\alpha$, say $\hat{\alpha}$, is given explicitly by

$$\hat{\alpha} = k/S_1(\hat{\beta}).$$

Theorem: The inequality for $1 \leqslant k \leqslant n$

$$\frac{2}{k} \sum_{i=1}^{k} t_i \sum_{j=1}^{n} t_j < \sum_{j=1}^{n} t_j^2 \tag{4}$$

is a sufficient condition which a (censored) sample from a mixed exponential $(\alpha, \beta)$ distribution must satisfy in order that maximum likelihood estimators of both parameters exist both positively and finitely. For proofs of the above theorems, see Myhre and Saunders [12].

## 6. COMPUTATIONAL CONSIDERATIONS

The question which now arises is: what kinds of samples will satisfy condition (4)? If $k = n$ we see (4) is equivalent with

$$\left(\frac{1}{n} \sum_{1}^{n} t_i\right)^2 < \frac{1}{n} \sum_{1}^{n} t_i^2 - \left(\frac{1}{n} \sum_{1}^{n} t_i\right)^2$$

from which we have the

Remark: A complete sample of failure times will satisfy (4) if the sample standard deviation exceeds the sample mean.

It can be shown that if $T$ has a mixed exponential $(\alpha, \beta)$ distribution then

$$E[T] = [\beta(\alpha - 1)]^{-1} \text{ for } \alpha > 1$$

$$\text{Var } [T] = \alpha[\beta^2(\alpha - 1)^2 (\alpha - 2)^2]^{-1} \text{ for } \alpha > 2.$$

Thus the standard deviation does exceed the mean for those values of the parameters where the mean $E[T]$ and the variance $V[T]$ exist.

Remark: A sample with $k < n$ failure times and the remaining $n - k$ observations truncated at $t_0$ will satisfy (4) if

$$t_0 > \eta_1\left[1 + \sqrt{\frac{2k}{n-k} + 1}\right] \simeq \eta_1 \frac{2n-k}{n-k} \text{ for } n \text{ large}$$

where $\eta_1 = (t_1 + \ldots + t_k)/k$ is the average failure time.

In the calculation of $\hat{\beta}$ from (iii) above, the equation, $C(\beta) = 0$, must be solved where $C(\beta) = \beta S_1'(\beta) - S_1(\beta)S_2(\beta)$ or

$$C(\beta) = \sum_{j=1}^{n} \frac{t_j\beta}{1 + t_j\beta} - \sum_{j=1}^{n} \ln(1 + t_j\beta) \sum_{i=1}^{k} \frac{1/k}{1 + t_i\beta}$$

where $t_1, \ldots, t_k$ are failure times and $t_{k+1}, \ldots, t_n$ are censored life times. We introduce notation for the sample moments as follows:

$$\eta_r = \frac{1}{k}\sum_{i=1}^{k} t_i^r,\ \zeta_r = \frac{1}{n}\sum_{j=1}^{n} t_j^r \text{ for } r = 1, 2, 3, \ldots, \tag{5}$$

then using the two expansions, valid for $|x| < 1$,

$$\ln(1 + x) = x - \frac{x^2}{2} + \frac{x^3}{3} - \ldots, \quad \frac{1}{1+x} = 1 - x + x^2 - \ldots$$

and substituting into $C$ and simplifying we find, upon neglecting terms of third order in $\beta$, that

$$(1 - \eta_1\beta + \eta_2\beta^2)\left(\zeta_1 - \zeta_2\frac{\beta}{2} + \zeta_3\frac{\beta^2}{3}\right) - [\zeta_1 - \zeta_2\beta + \zeta_3\beta^2] = 0.$$

Multiplying the first two together and collecting terms yields

$$\left(\frac{\zeta_2}{2} - \eta_1\zeta_1\right)\beta - \left(\frac{2}{3}\zeta_3 - \eta_2\zeta_1 - \frac{\eta_1\zeta_2}{2}\right)\beta^2 = 0.$$

We now notice that the condition equation (4), can be written in the notation of (5) as $\zeta_2 > 2\eta_1\zeta_1$.

Thus our computational procedure to decide upon the parametric representation of the distribution governing the observations which have been obtained is contained in the following.

Algorithm: Given $t_1, \ldots, t_k$ as failure times and $t_{k+1}, \ldots, t_n$ as censored times from a mixed exponential $(\alpha, \beta)$ distribution

(i) Compute the sample moments $\eta_1, \eta_2, \zeta_1, \zeta_2, \zeta_3$.
(ii) If $\zeta_2 < 2\eta_1\zeta_1$, assume observations from a constant failure rate distribution and estimate $\lambda$ by

$$\hat{\lambda} = \frac{k}{n\zeta_1}.$$

(iii) If $\zeta_2 > 2\eta_1\zeta_1$, assume observations are from a mixed exponential distribution and compute

$$\beta_0 = \frac{\zeta_2 - 2\eta_1\zeta_1}{\frac{4}{3}\zeta_3 - 2\eta_2\zeta_1 - \eta_1\zeta_2}$$

then use the Newton-Raphson iteration procedure, namely for $n = 0, 1, 2, \ldots .$

$$\beta_{n+1} = \beta_n - \frac{C(\beta_n)}{C'(\beta_n)}, \; \hat{\beta} = \lim_{n \to \infty} \beta_n, \text{ and}$$

$$\hat{\alpha} = \frac{k}{\sum_{j=1}^{n} \ln(1 + t_j\hat{\beta})}.$$

Practical experience indicates that the iteration converges very rapidly. Since the functions are very simple a small programmable electronic calculator, such as the HP-65, can be used to obtain these estimates. Programs for the HP-65 and HP-97 are available from the authors.

## 7. CONCLUSION

If a component has a life distribution with an increasing failure rate, the information necessary to estimate its parameters must contain failure times. In practice this means that virtually no observed failures, within a fleet of operational components, provide little information with which to assess reliability.

If a component has a constant failure rate then both failure times and alive times contribute equally to its estimation. The preceding study suggests that if a component has a life distribution with decreasing failure rate, it is the alive times within the data which contribute principally to the estimation of the parameters.

The problem of obtaining the usual sampling distributions of the maximum likelihood estimators of the parameters for the decreasing failure rate model studied seems to be difficult because the estimates are only implicitly defined. We have shown, however, that when they exist the m.l.e.'s for $\alpha$ and $\beta$, based on type I or on random censoring, are asymptotically normally distributed. We have also shown [11] that the distribution function estimated using the joint m.l.e.'s of the parameters is surprisingly closer to the true distribution for regions of interest in reliability theory, than is the estimated distribution function using a BLUE-k estimate for the scale parameter and a known shape parameter.

## REFERENCES

Afanas'ev N.N. (1940). Statistical Theory of the Fatigue Strength of Metals. *Zhurnal Tekhnicheska Fiziki, 10,* 1553-1568.

Barlow, R.E. and Proschan, Frank (1975). *Statistical Theory of Reliability and Life Testing.* Holt, Rinehart and Winston, Inc., New York.

Bhaltacharya, N. (1963). A Property of the Pareto Distribution. *Sankhya, Series B., 25,*, 195-196.

Birnbaum, Z.W. and Saunders, Sam C. (1969). A New Family of Life Distributions. *J. of Applied Prob., 6,* 319-327.

Cozzolino, John M. (1968). Probabilistic Models of Decreasing Failure Rate Processes. *Naval Research Logistic Quarterly, 15,* 361-374.

Epstein, B. and Sobel, M. (1953). Life Testing. *J. Amer. Statist. Assoc., 48,* 486-502.

Harris, C.M. and Singpurwalla, N.D. (1968). Life Distributions Derived from Stochastic Hazard Functions. *IEEE Trans. in Reliability, R-17,* 70-79.

Harris, C.M. and Singpurwalla, N.D. (1969). On Estimation in Weibull Distributions with Random Scale Parameters. *Naval Research Logistic Quarterly, 16,* 405-410.

Kulldorff, G. and Vännman, K. (1973). Estimation of the Location and Scale Parameters of a Pareto Distribution. *J. Amer. Statist. Assoc., 68,* 218-227.

Lomax, K.S. (1954). Business Failures: Another Example of the Analysis of Failure Data. *J. Amer. Statist. Assoc. 49,* 847-852.

Lucke, J. and Myhre, J. (1980). A Comparison Between MLE and BLUE for a Decreasing Failure Rate Distribution When Using Censored Data. Unpublished report, Institute of Decision Science, Claremont McKenna College.

Myhre, J. and Saunders, S. (1981). On Problems of Estimation for Two Parameter Decreasing Failure Rate Distributions Applied to Reliability. Unpublished report, Institute of Decision Science, Claremont McKenna College.

Proschan, F. (1963). Theoretical Explanation of Observed Decreasing Failure Rate. *Technometrics, 5,* 375-383.

Sunjata, M.H. (1974). *Sensitivity Analysis of a Reliability Estimation Procedure for a Component whose Failure Density is a Mixture of Exponential Failure Densities.* Naval Postgraduate School, Monterey. Unpublished thesis.

Vännman, K. (1976). Estimators Based on Order Statistics from a Pareto Distribution. *J. Amer. Statistic. Assoc., 71,* 704-707.

Weibull, W. (1961). *Fatigue Testing and Analysis of Results.* Pergaman Press, New York.

# Nonparametric Estimates for Reliability Growth

Carol Feltz and Richard Dykstra

*University of Missouri - Columbia*

**Abstract**

An important concept in the development of hardware systems is that of "reliability growth." In many situations it is reasonable (or at least desirable) to expect that a system should become more reliable as the system passes through various stages of development. One way of expressing this is to require the survival function of a system (the probability the lifetime of the system exceeds $t$ as a function of $t$) be increasing (nondecreasing) for each fixed $t$ as the system is refined and developed.

A method for obtaining good, nonparametric estimates (maximum likelihood) for survival functions subject to reliability growth restrictions have been found. It is hoped that these estimates may lead to good testing procedures for detecting patterns involving reliability growth. In particular, the ability to determine whether reliability development is "on schedule" could prove to be a useful statistical tool.

## 1 INTRODUCTION

Let us define the survival function $P(t)$ of an item as the probability the item lasts longer that $t$ units of time. Equivalently, if $X$ denotes the lifetime of the item,

$$P(t) = \text{Prob } (X > t).$$

Since $P(t)$ determines all the stochastic behavior of $X$ such as its mean value, median value, standard deviation, etc., we would like to know $P(t)$. By observing lifetimes and censored lifetimes of an independent sample of such items, we may hope to learn about the survival function $P(t)$.

If we have reason to believe that $P(t)$ is of a certain parametric form, estimating $P(t)$ is equivalent to estimating a parameter (or vector of parameters). If, however, $P(t)$ is not of that parametric form, error is present which cannot be improved upon by larger samples.

Thus nonparametric estimation may have some appeal if there is no compelling reason to believe that $P(t)$ is of any particular parametric form. The method of Maximum Likelihood Estimation (MLE) is viable in this setting and leads to the renowned Kaplan-Meier (1958) estimate of the survival function. (More on this later.)

If an item goes through $c$ stages of improvement, it is reasonable to assume that there are separate survival functions for each stage which satisfy

$$P_i(t) \geqslant P_{i-1}(t) \quad \text{for all} \quad t, \; i = 2, \ldots, c \tag{1.1}$$

(if improvement is really occuring at each stage).

However, ordinary Kaplan-Meier estimates may not satisfy (1.1) because of the inherent variability of the sampled lifetimes. Thus these estimates could be improved by imposing the order restrictions (1.1). (The restricted estimates may also prove useful in devising tests to determine whether improvement has really taken place.) Thus we wish to consider the problem of finding nonparametric MLE's of survival functions satisfying (1.1). (We will assume also that the survival functions correspond to nonnegative discrete distributions.)

## 2 NOTATION

To indicate the general form of the likelihood function, we consider an independent sample of observations (possibly censored) from the $i$th stage of development. We assume we have complete observations (failures) on a subset of the times $S_1 < S_2 < \ldots < S_m$ $(S_0 = 0,\ S_{m+1} = \infty)$. We let $\delta_j^{(i)}$ denote the number of failures at $S_j$ and let $\lambda_j^{(i)}$ denote the number of censored observations (losses) in the interval $[S_j, S_{j+1})$ (at times $L_{1j}, L_{2j}, \ldots, L\lambda_j^{(i)},j$). Thus (assuming discreteness for $P_i(t)$) the likelihood function corresponding to the $i$th stage is given by

$$L(P_i(t)) = \prod_{l=1}^{\lambda_0^{(i)}} P_i(L_{l,0}) \cdot \prod_{j=1}^{m} \{[P_i(S_j - 0) - P_i(S_j)]^{\delta_j^{(i)}} \prod_{l=1}^{\lambda_j^{(i)}} P_i(L_{l,j})\} \tag{2.1}$$

(We assume that our censoring times are fixed, but the same results will obtain if they are random variables independent of the failure times.)

Thus our problem is to find survival functions $P_i(t)$, $i = 1, \ldots, c$ which maximize the total likelihood

$$L(P_1(t), P_2(t), \ldots, P_c(t)) = \prod_{i=1}^{c} L(P_i(t))$$

and satisfy the order restrictions (1.1). (A survival function is nonincreasing, right-continuous, converging to 0 at $\infty$ and converging to 1 at $-\infty$.)

For arbitrary survival functions $P_1(t), \ldots, P_c(t)$, we will not create any new violations of (1.1) and will not decrease our likelihood if we assume that $P_c(t)$ is a right continuous step function with jumps at $S_i$ where $P_c(S_i)$ is left unchanged. Thus we may assume that our MLE $\hat{P}_c(t)$ is a step function with jumps among the $S_i$. Thus we may assume that

$$L(P_c(t)) = \prod_{j=1}^{m} [P_c(S_{j-1}) - P_c(S_j)]^{\delta_j^{(c)}} P_c(S_j)^{\lambda_j^{(c)}} .$$

(Throughout this paper we shall treat $-\infty$ and $+\infty$ as real numbers less than and greater than all other real numbers, respectively. We shall also adopt the conventions that $\ln(0) = -\infty$, $\infty/\infty = -\infty/-\infty = 1$, $-\infty/\infty = \infty/-\infty = -1$, $0(\pm\infty) = 0$, $0^\circ = 1$, and $0/0 = 1$ unless otherwise stated.)

Once we know that $\hat{P}_c(t)$ is of the above form, it follows that $\hat{P}_{c-1}(t)$ must also be a right continuous step function wth jumps occurring among $S_1, \ldots, S_m$. By induction, the same is true for $P_1(t), P_2(t), \ldots, P_{c-2}(t)$. Thus maximizing the likelihood over all discrete survival functions is equivalent to maximizing it over those with jumps occurring among $S_1, \ldots, S_m$ where $S_1 < S_2 < \ldots < S_m$ denote the ordered failure times of all $c$ combined samples. Thus it will suffice to maximize

$$L(P_1, P_2, \ldots, P_c) = \prod_{i=1}^{c} \prod_{j=1}^{m} [P_i(S_{j-1}) - P_i(S_j)]^{\delta_j^{(i)}} P_i(S_j)^{\lambda_j^{(i)}}$$

subject to

$$P_c(S_j) \geqslant P_{c-1}(S_j) \geqslant \ldots \geqslant P_1(S_j),\ j = 1, \ldots, m.$$

(Of course we also have the implied constraints

$$1 \geqslant P_i(S_1) \geqslant P_i(S_2) \geqslant \ldots \geqslant P_i(S_m) \geqslant 0,\ i = 1, \ldots, c.)$$

Equivalently, we wish to maximize

$$\prod_{i=1}^{c} \prod_{j=1}^{m} \left[1 - \frac{P_i(S_j)}{P_i(S_{j-1})}\right]^{\delta_j^{(i)}} \left[\frac{P_i(S_{j-1})}{P_i(S_{j-2})} \cdots P_i(S_1)\right]^{\delta_j^{(i)}} \left[\frac{P_i(S_j)}{P_i(S_{j-1})} \cdots P(S_1)\right]^{\lambda_j^{(i)}},$$

or letting $\tilde{p}_j^{(i)} = \dfrac{P_i(S_j)}{P_i(S_{j-1})}$, to maximize

$$\prod_{i=1}^{c} \prod_{j=1}^{m} (1 - \tilde{p}_j^{(i)})^{\delta_j^{(i)}} \tilde{p}_j^{(i)\lambda_j^{(i)}} \prod_{l<j} \tilde{p}_l^{(i)^{\lambda_j^{(i)}+\delta_j^{(i)}}} \tag{2.2}$$

subject to $\prod_{l=1}^{j} \tilde{p}_l^{(i)} \geqslant \prod_{l=1}^{j} \tilde{p}_l^{(i-1)}$, $1 \geqslant \tilde{p}_j^{(i)} \geqslant 0$ for all $i, j$. If we let $n_j^{(i)} = \sum_{l=j}^{m} (\delta_l^{(i)} + \lambda_l^{(i)})$ denote the number of items in the $i$th sample surviving just prior to $S_j$, (2.2) becomes

$$\prod_{i=1}^{c} \prod_{j=1}^{m} (1 - \tilde{p}_j^{(i)})^{\delta_j^{(i)}} \tilde{p}_j^{(i)^{n_j^{(i)} - \delta_j^{(i)}}}.$$

Finally, making the change of variables

$$p_j^{(i)} = \ln \tilde{p}_j^{(i)} \quad \text{for all } j, i,$$

and considering the natural log of the likelihood, our problem is to maximize

$$\sum_{i=1}^{c} \sum_{j=1}^{m} \delta_j^{(i)} \ln (1 - e^{p_j^{(i)}}) + (n_j^{(i)} - \delta_j^{(i)}) p_j^{(i)} \tag{2.3}$$

subject to the constraints

$$\sum_{l=1}^{j} p_l^{(i)} \geqslant \sum_{l=1}^{j} p_l^{(i-1)}, \; 0 \geqslant p_j^{(i)} \geqslant -\infty \tag{2.4}$$

for all $i$, $j$. It is important that (2.3) is concave, and that the constraints in (2.4) are linear. Note also that if $\delta_j^{(i)} = 0$, (2.3) is linear in $p_j^{(i)}$ and that the coefficient $n_j^{(i)} - \delta_j^{(i)} \geqslant 0$. Thus in this case we wish to make $p_j^{(i)}$ as large as possible subject to the restrictions in (2.4).

## 3 SOLUTION

Since we have already established that our MLE's $\hat{P}_i(t)$ are right-continuous step functions with jumps occurring among the combined failure times $S_1 < S_2 < \ldots < S_m$, it will suffice to find $\hat{P}_i(S_j)$ for all $i$, $j$. By the invariance property of MLE's, however,

$$\hat{P}_i(t) = \exp \left[ \sum_{j; S_j \leqslant t} \hat{p}_j^{(i)} \right]$$

where $\hat{p}_j^{(i)}$ maximizes (2.3) subject to the restrictions in (2.4). Hence we will concern ourselves only with the $\hat{p}_j^{(i)}$'s.

If we impose only the restrictions $0 \geqslant p_j^{(i)} \geqslant -\infty$ in (2.4), each term in the sum (2.3) may be maximized separately to obtain

$$\hat{p}_j^{(i)} = \ln \left[ \frac{n_j^{(i)} - \delta_j^{(i)}}{n_j^{(i)}} \right] \quad \text{for all } i, j$$

which yield the well-known Kaplan-Meier estimates.

If $c = 2$, and all $\lambda_j^{(i)} = 0$ (no censored observations are present), expressions for $\hat{P}_1$ and $\hat{P}_2$ have been found by Brunk, Franck, Hanson, and Hogg (1966).

If $c = 2$ and the $\lambda_j^{(i)}$ are not necessarily zero, Dykstra (1980) has found expressions for the $\hat{p}_j^{(i)}$ which bear an interesting relationship to the Kaplan-Meier estimates. In particular,

$$\hat{p}_j^{(2)} = \ln\left[\frac{n_j^{(2)} + k_j - \delta_j^{(2)}}{n_j^{(2)} + k_j}\right] \quad \text{and} \tag{3.1}$$

$$\hat{p}_j^{(1)} = \ln\left[\frac{n_j^{(1)} - k_j - \delta_j^{(1)}}{n_j^{(1)} - k_j}\right] \tag{3.2}$$

for correct values of of the $k_j$. (Note that the solution is still of the Kaplan-Meier form if one acts as if $k_j$ items of the first sample which had survived to just prior to time $S_j$ were really items from the second sample which had survived to just prior to time $S_j$.)

The $k_j$ may be characterized as follows. If $\delta_1^{(1)} = \ldots = \delta_{j_0}^{(1)} = 0$, $\delta_{j_0+1}^{(1)} > 0$, then $k_1 = n_1^{(1)}$, $k_2 = n_2^{(1)}$, $\ldots$, $k_{j_0} = n_{j_0}^{(1)}$, and $\hat{p}_j^{(1)} = \ln(0/0)$ is taken to be

$$\hat{p}_j^{(2)} = \ln\left[\frac{n_j^{(2)} + k_j - \delta_j^{(2)}}{n_j^{(2)} + k_j}\right]$$

for $j = 1, \ldots, j_0$. Then $k_{j_0+1}, \ldots, k_m$ are the unique values such that

(i) $\hat{p}_j^{(1)}, \hat{p}_j^{(2)}$ (as defined in (3.1) and (3.2)) satisfy the constraints in (2.4),

(ii) $k_1 \geqslant k_2 \geqslant \ldots \geqslant 0$, and (3.3)

(iii) whenever $k_j > k_{j+1}$, $\sum_{l=1}^{j} \hat{p}_l^{(2)} = \sum_{l=1}^{j} \hat{p}_l^{(1)}$.

From this characterization of the $k_j$'s, we may find other ways of expressing them. For example, let $k_{a,b}$ denote the constant $k$ such that

$$\sum_{j=a}^{b} \ln\left\{\frac{n_j^{(1)} - k - \delta_j^{(1)}}{n_j^{(1)} - k}\right\} = \sum_{j=a}^{b} \ln\left\{\frac{n_j^{(2)} + k - \delta_j^{(2)}}{n_j^{(2)} + k}\right\} \tag{3.4}$$

if $k \geqslant 0$ and exists, and 0 otherwise ($a > j_0$). Then

$$k_i = \min_{j_0 < a \leqslant i} \max_{i \leqslant b} k_{a,b} \text{ for } i = j_0 + 1, \ldots, m.$$

The easiest way to compute the $k_i$'s is probably the following:

1. Find the largest $i_l > j_0$ such that the $k_{i_l} > 0$ which solves

$$\sum_{j=j_0+1}^{i_1} \ln\left[\frac{n_j^{(1)} - k - \delta_j^{(1)}}{n_j^{(1)} - k}\right] = \sum_{j=j_0+1}^{i_1} \ln\left[\frac{n_j^{(2)} + k - \delta_j^{(2)}}{n_j^{(2)} + k}\right]$$

is maximized. Then $k_i = k_{i_l}$ for $j_0 < i \leqslant i_l$.

2. Then find the largest $i_2 > i_1$ such that the $k_{i_2} > 0$ which solves

$$\sum_{j=i_1+1}^{i_2} \ln\left(\frac{n_j^{(1)} - k - \delta_j^{(1)}}{n_j^{(1)} - k}\right) = \sum_{j=i_1+1}^{i_2} \ln\left(\frac{n_j^{(2)} + k - \delta_j^{(2)}}{n_j^{(2)} + k}\right)$$

is maximized. Then $k_i = k_{i_2}$ for $i_1 < i \leqslant i_2$.

3. Etc. If at some point, no such positive number $k_{i_l}$ exists, then

$$k_i = 0 \text{ for } i > i_{l-1} \,.$$

Interestingly, the case $c > 2$, while much more difficult, has an analogous solution which should appeal to our heuristic instincts. In particular, with two samples, the $k_i$'s can be interpreted as how many unfailed items must be transferred from the first sample to the second. In the more general case specified in (1.2), we will need a vector of values $\underset{\sim}{k}^{(i)}, i = 1, \ldots, c-1$, to indicate how many unfailed items need to be transferred from the $i$th sample to the $(i+1)^{\text{st}}$ sample. The general form of the solutions can be expressed as

$$\hat{p}_j^{(1)} = \ln\left(\frac{n_j^{(1)} - k_j^{(1)} - \delta_j^{(1)}}{n_j^{(1)} - k_j^{(1)}}\right),$$

$$\vdots$$

$$\hat{p}_j^{(i)} = \ln\left(\frac{n_j^{(i)} + k_j^{(i-1)} - k_j^{(i)} - \delta_j^{(i)}}{n_j^{(i)} + k_j^{(i-1)} - k_j^{(i)}}\right), \quad i = 2, \ldots, c-1 \tag{3.5}$$

$$\vdots$$

$$\hat{p}_j^{(c)} = \ln\left(\frac{n_j^{(c)} + k_j^{(c-1)} - \delta_j^{(c)}}{n_j^{(c)} + k_j^{(c-1)}}\right),$$

with certain values specified for some 0/0 cases.

The difficult problem is that of determining the values of the $k_j^{(i)}$'s. When certain of the $\delta_j^{(i)}$'s are zero, corresponding $p_j^{(i)}$'s must be of a certain form.

If $\delta_j^{(i)} = 0$ for $j \leqslant j_0$ and $i \leqslant i_0$, and $\delta_{j_0}^{(i_0+1)} > 0$, then

$$\hat{p}_{j_0}^{(1)} = \hat{p}_{j_0}^{(2)} = \ldots = \hat{p}_{j_0}^{(i_0+1)} = \frac{\left(\sum_{i=1}^{i_0+1} n_{j_0}^{(i)}\right) - k_{j_0}^{(i_0+1)} - \delta_{j0}^{(i_0+1)}}{\sum_{i=1}^{i_0+1} n_{j_0}^{(i)} - k_{j_0}^{(i_0+1)}} \,. \tag{3.6}$$

(Note that this is consistent with (3.5) if $k_{j_0}^{(i)} = \sum_{l=1}^{i} n_{j_0}^{(l)}$, $i = 1, \ldots, i_0$, and 0/0 is defined appropriately.)

We can characterize the solution to (2.3) under the constraints (2.4) as follows:

THEOREM 2.1. The solution to (2.3) under the constraints (2.4) is given by (3.5) and (3.6) when the $k_j^{(i)}$ in (3.5) and (3.6) are such that

(i) the $\hat{p}_j^{(i)}$ in (3.5) and (3.6) satisfy the constraints in (2.4),
(ii) $k_1^{(i)} \geqslant k_2^{(i)} \geqslant \ldots \geqslant k_m^{(i)} \geqslant 0$ for $i = 1, 2, \ldots, c-1$,
(iii) whenever $k_j^{(i)} > k_{j+l}^{(i)}$, $\sum_{l=1}^{j} \hat{p}_l^{(i+1)} = \sum_{l=1}^{j} \hat{p}_l^{(i)}$.

Unfortunately, for $c > 2$ there is not a practical straight forward way that we have found of finding the $k_j^{(i)}$'s as there is in the case $c = 2$. However, we have been able to find an iterative technique which depends only upon the ability to find the $k_j^{(i)}$ when $c = 2$ (which is straightforward) which converges to the correct solution quite rapidly.

This algorithm is described as follows:

1. Find $k_j^{(c-1)}$ (1) in (3.1)-(3.2) when comparing *only* the $c^{\text{th}}$ and $(c-1)^{\text{st}}$ samples.
2. Replace $n_j^{(c)}$ by $n_j^{(c)} + k_j^{(c-1)}$ (1) and $n_j^{(c-1)}$ by $n_j^{(c-1)} - k_j^{(c-1)}(1)$.
3. Find $k_j^{c-2}(1)$ in (3.1)-(3.2) when comparing *only* the $(c-1)^{\text{st}}$ and $(c-2)^{\text{nd}}$ samples.
4. Replace $n_j^{(c-1)} - k_j^{(c-1)}$ (1) by $n_j^{(c-1)} - k_j^{(c-1)}$ (1) $+ k_j^{(c-2)}(1)$ and $n_j^{(c-2)}$ by $n_j^{(c-2)} - k_j^{(c-2)}(1)$.
5. Proceed as in steps 3 and 4 until MLE's for $1^{\text{st}}$ and $2^{\text{nd}}$ samples have been obtained and $n_j^{(2)} - k_j^{(2)}(1)$ has been replaced by $n_j^{(2)} - k_j^{(2)}(1) + k_j^{(1)}$ (1) and $n_j^{(1)}$ has been replaced by $n_j^{(1)} - k_j^{(1)}(1)$.
6. Find $k_j^{(c-1)}(2)$ in (3.1)-(3.2) when comparing only the $c^{\text{th}}$ and $(c-1)^{\text{st}}$ samples *and* $k_j^{(c-1)}(1)$ *has been set equal to zero.* Then replace $k_j^{(c-1)}$ (1) by $k_j^{(c-1)}$ (2) in step 2.
7. Etc.

By using this algorithm,

$$k_j^{(i)}(l) \rightarrow k_j^{(i)} \quad \text{as} \quad l \rightarrow \infty$$

where $k_j^{(i)}$ are the required constants in (3.5) and (3.6).

## 4 AN EXAMPLE

To illustrate our estimates, we consider some data presented in a paper by Nel-

**Test Voltage**

| 30 KV | 32 KV | 34 KV | 36 KV |
|---|---|---|---|
| 7.74 | .27 | .19 | .35 |
| 17.05 | .40 | .78 | .59 |
| 20.46 | .69 | .96 | .96 |
| 21.02 | .79 | 1.31 | .99 |
| 22.66 | 2.75 | 2.78 | 1.69 |
| 43.40 | 3.91 | 3.16 | 1.97 |
| 47.30 | 9.88 | 4.15 | 2.07 |
| 139.07 | 13.95 | 4.67 | 2.58 |
| 144.12 | 15.93 | 4.85 | 2.71 |
| 175.88 | 27.80 | 6.50 | 2.90 |
| 215.10 | 53.24 | 7.35 | 3.67 |
| | 82.85 | 8.01 | 3.99 |
| | 89.29 | 8.27 | 5.35 |
| | 100.58 | 12.06 | 13.77 |
| | 194.90 | 31.75 | 25.50 |
| | | 32.52 | |
| | | 33.91 | |
| | | 36.71 | |
| | | 72.89 | |

**Figure 1 Time to breakdown of an insulating fluid subjected to various constant elevated test voltage.**

son (1972) concerning times to breakdown of an insulating fluid subjected to various constant elevated voltages. This data is presented in Fig. 1. Since this data is obtained from an accelerated life testing situation, it is intuitively obvious that stochastic orderings for the survival functions corresponding to various voltage levels should hold. (The higher the voltage, the lower the survival function.) However the unrestricted MLE's do not obey these stochastic orderings as shown in Figure 2.

To indicate how quickly the convergence of the $k_j^{(i)}(l)$ take place as $l \to \infty$, we list the first two iterations. To conform to the notation of the paper, we let $P_4$ indicate the survival function for a test voltage of 30 KV, $P_3$ for 32 KV, $P_2$ for 34 KV, and $P_1$ for 36 KV.

First Iteration.
$$k_j^{(3)}(1) = 0 \text{ for all } j$$

$$k_j^{(2)}(1) = \begin{cases} 21.7503, & j \leqslant 6 \\ 3.0000, & 7 \leqslant j \leqslant 8 \\ 0, & 9 \leqslant j \end{cases}$$

$$k_j^{(1)}(1) = \begin{cases} 15.0000, & j \leqslant 2 \\ 0, & 3 \leqslant j \end{cases}$$

| Time | P4 30 KV | P3 32 KV | P2 34 KV | P1 36 KV |
|---|---|---|---|---|
| .19 | 1.000000 | 1.000000 | .947368 | 1.000000 |
| .27 | 1.000000 | .933333 | .947368 | 1.000000 |
| .35 | 1.000000 | .933333 | .947368 | .933333 |
| .40 | 1.000000 | .866667 | .947368 | .933333 |
| .59 | 1.000000 | .866667 | .947368 | .866667 |
| .69 | 1.000000 | .800000 | .847368 | .866667 |
| .78 | 1.000000 | .800000 | .947368 | .866667 |
| .79 | 1.000000 | .733333 | .894737 | .866667 |
| .96 | 1.000000 | .733333 | .842105 | .800000 |
| .99 | 1.000000 | .733333 | .842105 | .733333 |
| 1.31 | 1.000000 | .733333 | .789474 | .733333 |
| 1.69 | 1.000000 | .733333 | .789474 | .666667 |
| 1.97 | 1.000000 | .733333 | .789474 | .600000 |
| 2.07 | 1.000000 | .733333 | .789474 | .533333 |
| 2.58 | 1.000000 | .733333 | .789474 | .466667 |
| 2.71 | 1.000000 | .733333 | .789474 | .400000 |
| 2.75 | 1.000000 | .666667 | .789474 | .400000 |
| 2.78 | 1.000000 | .666667 | .736842 | .400000 |
| 2.90 | 1.000000 | .666667 | .736842 | .333333 |
| 3.16 | 1.000000 | .666667 | .684211 | .333333 |
| 3.67 | 1.000000 | .666667 | .684211 | .266667 |
| 3.91 | 1.000000 | .600000 | .684211 | .266667 |
| 3.99 | 1.000000 | .600000 | .684211 | .200000 |
| 4.15 | 1.000000 | .600000 | .631579 | .200000 |
| 4.67 | 1.000000 | .600000 | .578947 | .200000 |
| 4.85 | 1.000000 | .600000 | .526316 | .200000 |
| 5.35 | 1.000000 | .600000 | .526316 | .133333 |
| 6.50 | 1.000000 | .600000 | .473684 | .133333 |
| 7.35 | 1.000000 | .600000 | .421053 | .133333 |
| 7.74 | .909091 | .600000 | .421053 | .133333 |
| 8.01 | .909091 | .600000 | .368421 | .133333 |
| 8.27 | .909091 | .600000 | .315789 | .133333 |
| 9.88 | .909091 | .533333 | .315789 | .133333 |
| 12.06 | .909091 | .533333 | .263158 | .133333 |
| 13.77 | .909091 | .533333 | .263158 | .066667 |
| 13.95 | .909091 | .466667 | .263158 | .066667 |
| 15.93 | .909091 | .400000 | .263158 | .066667 |
| 17.05 | .818182 | .400000 | .263158 | .066667 |
| 20.46 | .727273 | .400000 | .263158 | .066667 |
| 21.02 | .636364 | .400000 | .263158 | .066667 |
| 22.66 | .545455 | .400000 | .263158 | .066667 |
| 25.50 | .545455 | .400000 | .263158 | .000000 |
| 27.80 | .545455 | .333333 | .263158 | .000000 |
| 31.75 | .545455 | .333333 | .210526 | .000000 |
| 32.52 | .545455 | .333333 | .157895 | .000000 |
| 33.91 | .545455 | .333333 | .105263 | .000000 |
| 36.71 | .545455 | .333333 | .052632 | .000000 |
| 43.40 | .454545 | .333333 | .052632 | .000000 |
| 47.30 | .363636 | .333333 | .052632 | .000000 |
| 53.24 | .363636 | .266667 | .052632 | .000000 |
| 72.89 | .363636 | .266667 | .000000 | .000000 |
| 82.85 | .363636 | .200000 | .000000 | .000000 |
| 89.29 | .363636 | .133333 | .000000 | .000000 |
| 139.07 | .272727 | .066667 | .000000 | .000000 |
| 144.12 | .181818 | .066667 | .000000 | .000000 |
| 175.88 | .090909 | .066667 | .000000 | .000000 |
| 194.90 | .090909 | .000000 | .000000 | .000000 |
| 215.10 | .000000 | .000000 | .000000 | .000000 |

**Figure 2 Unrestricted maximum likelihood (Kaplan-Meier) estimates of survival functions.**

| Time | P4 30 KV | P3 32 KV | P2 34 KV | P1 36 KV |
|---|---|---|---|---|
| .19 | 1.000000 | 1.000000 | .917937 | .917937 |
| .27 | 1.000000 | .972647 | .917937 | .917937 |
| .35 | 1.000000 | .972647 | .917937 | .856741 |
| .40 | 1.000000 | .945294 | .917937 | .856741 |
| .59 | 1.000000 | .945294 | .917937 | .795545 |
| .69 | 1.000000 | .917940 | .917937 | .795545 |
| .78 | 1.000000 | .917940 | .856215 | .795545 |
| .79 | 1.000000 | .856219 | .856215 | .795545 |
| .96 | 1.000000 | .856219 | .805850 | .734349 |
| .99 | 1.000000 | .856219 | .805850 | .673154 |
| 1.31 | 1.000000 | .856219 | .755484 | .673154 |
| 1.69 | 1.000000 | .856219 | .755484 | .611958 |
| 1.97 | 1.000000 | .856219 | .755484 | .550762 |
| 2.07 | 1.000000 | .856219 | .755484 | .489566 |
| 2.58 | 1.000000 | .856219 | .755484 | .428370 |
| 2.71 | 1.000000 | .856219 | .755484 | .367175 |
| 2.75 | 1.000000 | .776530 | .755484 | .367175 |
| 2.78 | 1.000000 | .776530 | .705118 | .367175 |
| 2.90 | 1.000000 | .776530 | .705118 | .305979 |
| 3.16 | 1.000000 | .776530 | .654753 | .305979 |
| 3.67 | 1.000000 | .776530 | .654753 | .244783 |
| 3.91 | 1.000000 | .696840 | .654753 | .244783 |
| 3.99 | 1.000000 | .696840 | .654753 | .183587 |
| 4.15 | 1.000000 | .696840 | .604387 | .183587 |
| 4.67 | 1.000000 | .696840 | .554021 | .183587 |
| 4.85 | 1.000000 | .696840 | .503656 | .183587 |
| 5.35 | 1.000000 | .696840 | .503656 | .122392 |
| 6.50 | 1.000000 | .696840 | .453290 | .122392 |
| 7.35 | 1.000000 | .696840 | .402925 | .122392 |
| 7.74 | .911155 | .696840 | .402925 | .122392 |
| 8.01 | .911155 | .696840 | .352560 | .122392 |
| 8.27 | .911155 | .696840 | .302194 | .122392 |
| 9.88 | .911155 | .617151 | .302194 | .122392 |
| 12.06 | .911155 | .617151 | .251828 | .122392 |
| 13.77 | .911155 | .617151 | .251828 | .061196 |
| 13.95 | .911155 | .537462 | .251828 | .061196 |
| 15.93 | .911155 | .457773 | .251828 | .061196 |
| 17.05 | .822309 | .457773 | .251828 | .061196 |
| 20.46 | .733463 | .457773 | .251828 | .061196 |
| 21.02 | .644618 | .457773 | .251828 | .061196 |
| 22.66 | .555773 | .457773 | .251828 | .061196 |
| 25.50 | .555773 | .457773 | .251828 | .000000 |
| 27.80 | .555773 | .378084 | .251828 | .000000 |
| 31.75 | .555773 | .378084 | .201462 | .000000 |
| 32.52 | .555773 | .378084 | .151097 | .000000 |
| 33.91 | .555773 | .378084 | .100731 | .000000 |
| 36.71 | .555773 | .378084 | .050366 | .000000 |
| 43.40 | .466928 | .378084 | .050366 | .000000 |
| 47.30 | .378083 | .378083 | .050366 | .000000 |
| 53.24 | .378083 | .302467 | .050366 | .000000 |
| 72.89 | .378083 | .302467 | .000000 | .000000 |
| 82.85 | .378083 | .226850 | .000000 | .000000 |
| 89.29 | .378083 | .151234 | .000000 | .000000 |
| 139.07 | .283562 | .075617 | .000000 | .000000 |
| 144.12 | .189041 | .075617 | .000000 | .000000 |
| 175.08 | .094521 | .075617 | .000000 | .000000 |
| 194.90 | .094521 | .000000 | .000000 | .000000 |
| 215.10 | .000000 | .000000 | .000000 | .000000 |

**Figure 3 Maximum likelihood estimates of survival functions under stochastic ordering restrictions P1 < P2 < P3 < P4.**

Second Iteration.

$$k_j^{(3)}(2) = \begin{cases} .2526, & j \leqslant 49 \\ 0, & 50 \leqslant j \end{cases}$$

$$k_j^{(2)}(2) = \begin{cases} 21.8136, & j \leqslant 6 \\ 3.1263, & 7 \leqslant j \leqslant 80 \\ 0, & 9 \leqslant j \end{cases}$$

$$k_j^{(1)}(2) = \begin{cases} 15.0000, & j \leqslant 2 \\ 0, & 3 \leqslant j \end{cases}$$

Limiting Values (MLE).

$$k_j^{(3)} = \begin{cases} .2555, & j \leqslant 49 \\ 0, & 50 \leqslant j \end{cases}$$

$$k_j^{(2)} = \begin{cases} 21.8143, & j \leqslant 6 \\ 3.1278, & 7 \leqslant j \leqslant 8 \\ 0, & 9 \leqslant j \end{cases}$$

$$k_j^{(1)} = \begin{cases} 15.0000, & j \leqslant 2 \\ 0, & 3 \leqslant j \end{cases}$$

The MLE's of the survival functions under the stochastic order restrictions given in (1.1) are then obtained from (3.5) and (3.6) using the limiting values of the $k_j^{(i)}$'s given in (4.1). These MLE's are tabled in Fig. 3.

## REFERENCES


Brunk, H.D., Franck, W.E., Hanson, D.L., and Hogg, R.V. (1966). Maximum likelihood estimation of the distributions of two stochastically ordered random variables. *Journal of American Statistical Assn.* 61, 1067-1080.

Dykstra, Richard L. (1980). Maximum likelihood estimation of the survival functions of stochastically ordered random variables. Univ. of Missouri, Columbia, Dept. of Statistics Technical Report No. 91. (To appear, *Journal of American Statistical Association.*)

Kaplan, E.L., and Meier, Paul (1958). Nonparametric estimation from incomplete observations. *Journal of American Statistical Assn.* 53, 457-481.

Nelson, Wayne (1972). Graphical analysis of accelerated life test data wth the inverse power law model. *IEEE Transactions on Reliability*, Vol. R-21, 2-11.

# Statistical Analyses for Nondestructive Testing

D. B. Owen

*Southern Methodist University*

**Abstract**

A statistical method is developed for making an inference about a performance variate based on an observation of a screening variate. This is the typical situation in nondestructive testing where to measure the variable of primary interest would destroy or degrade the item under study. Typically the performance variable is lifetime. The nondestructive testing engineer must look at other related variables and based upon some mathematical analysis and engineering judgments, decide if the item can meet the minimum lifetime requirements.

The method described in this paper allows the engineer to make this inference from the screening variable to the performance variable based on a training set. That is, data are gathered on the screening variable and the performance variable for a set of $n$ items. Then all future items are screened according to the rule developed and among the accepted items there is a preassigned probability that at least a given proportion of the items will have minimum lifetimes. The method is simple and easy to apply.

## 1 INTRODUCTION

In the typical nondestructive testing situation measurements are made on one random variable while the inference is to a second random variable. For example, a structural part of an aircraft may be X-rayed or measured using ultrasound devices. Based on those measurements we wish to have some high assurance that the aircraft part will last at least for some preassigned length of time.

Until recently the inference has been made based on the presence or absence of any cracks at all showing in the X-ray. However, X-ray technology has now advanced to the point where cracks are shown which may not have any significant effect on the desired lifetime of the item on test. Packman, et al. (1976) point out the need for statistical methods in handling problems of this type.

A nondestructive testing engineer evaluates the X-ray photographs and based upon his best engineering judgment and what calculations he can make decides whether the structure will last for the required length of time. The method described here allows him to make the inference from the X-rays to the lifetime based on a statistical model. Each item is subjected to a test and some composite measure of the X-ray density is developed which is then correlated with the performance variable (lifetime) which has the specification placed on it. Hence the model takes into account the engineer's judgment through the method of developing the screening variable over a training set; and then the process can be automated. In this way human errors are eliminated from the process. The technique also leads to greater consistency in the judgments that are applied.

Let us be clear at this point that development of this measurement to represent the X-ray will not be trivial. It will take a great deal of cooperative effort between engineers and statisticians to develop meaningful and consistent estimators. Each application will probably require a new development effort.

I do not want to underestimate the effort that will be required to do this properly in each case. On the other hand, in this paper I will assume that the X-measurement (on the correlated variate) and the *Y*-measurement (on the performance variate), have been developed and are jointly bivariate normally distributed.

In mathematical terms we want to be, say, 99% sure that the remaining lifetime of the part is at least $L$ flying hours.

The technique which will be discussed develops the statistical model for this problem and gives the procedure which must be followed to arrive at the assurance which is sought.

Table 1 gives, for example, the increase in the proportion of aircraft meeting our criterion (in our example, at least $L$ flying hours) if we start off with 70% of the aircraft meeting the criterion and we select a proportion (selection ratio) with the highest scores on their X-rays where the correlation between the X-rays and the lifetime is indicated in the left column. For example, if we choose those aircraft which are in the upper 40% in X-ray scores and we have a correlation 0.75 we will increase our percentage of aircraft with the needed flying time from 70% to 95%. Of course, the use in this context would usually be for special missions. However, the technique which has been developed here is much more general than that.

Now let me show you Table 2 where you can use two criteria to select the aircraft for the special mission. Here the subscript one (1) refers to the performance variable, (lifetime) and the subscripts two and three (2 and 3) refer to the screening variables. This time you may want to think of the second variable as some measure of the structure of the aircraft and the third variable as a measure of the engine viability. The first variable is still the lifetime of the aircraft and we consider that we are successful when the lifetime is at least $L$ flying hours.

**Table 1 Proportion meeting requirements after screening δ when proportion meeting requirements before Screening is γ = 0.70**

| Correlation | Selection Range β | | | | | | |
|---|---|---|---|---|---|---|---|
| ρ | 0.10 | 0.20 | 0.30 | 0.40 | 0.50 | 0.60 | 0.70 |
| .00 | 0.70 | 0.70 | 0.70 | 0.70 | 0.70 | 0.70 | 0.70 |
| .10 | 0.76 | 0.75 | 0.74 | 0.73 | 0.73 | 0.72 | 0.72 |
| .15 | 0.79 | 0.77 | 0.76 | 0.75 | 0.74 | 0.73 | 0.73 |
| .20 | 0.81 | 0.79 | 0.78 | 0.77 | 0.76 | 0.75 | 0.74 |
| .25 | 0.84 | 0.81 | 0.80 | 0.78 | 0.77 | 0.76 | 0.75 |
| .30 | 0.86 | 0.84 | 0.82 | 0.80 | 0.78 | 0.77 | 0.75 |
| .35 | 0.89 | 0.86 | 0.83 | 0.82 | 0.80 | 0.78 | 0.76 |
| .40 | 0.91 | 0.88 | 0.85 | 0.83 | 0.81 | 0.79 | 0.77 |
| .45 | 0.93 | 0.90 | 0.87 | 0.85 | 0.83 | 0.81 | 0.78 |
| .50 | 0.94 | 0.91 | 0.89 | 0.87 | 0.84 | 0.82 | 0.80 |
| .60 | 0.97 | 0.95 | 0.92 | 0.90 | 0.87 | 0.85 | 0.82 |
| .70 | 0.99 | 0.97 | 0.96 | 0.93 | 0.91 | 0.88 | 0.84 |
| .75 | 1.00 | 0.98 | 0.97 | 0.95 | 0.92 | 0.89 | 0.86 |

If the proportion meeting the criterion of at least $L$ flying hours is 0.70 before selection and if the correlation between flying hours and structural strength is $\rho_{12} = 0.5$; correlation between flying hours and engine viability is $\rho_{13} = 0.4$, and the correlation between structural strength and engine viability is $\rho_{23} = 0.1$, then selecting the 40% of all aircraft with 62.1% in the high structural strength and 62.1% in the high engine viability categories will raise the success of the mission from 70% to 88.2%. The technique of having two screening variables is especially important when we have a minimum requirement for two screening variables, as in our example, which cannot be combined into a single variate. In other words for most missions it would *not* be desirable to allow an extra high structural measurement to offset a low engine measurement.

**Table 2 Proportion meeting requirements after screening δ when proportion meeting requirements before screening is γ = 0.70**

| | | | True Selection Ratio | | | | | | | |
|---|---|---|---|---|---|---|---|---|---|---|
| | | | .1 | .2 | .3 | .4 | .5 | .6 | .7 | .8 |
| | | | β(.297) | (.430) | (.533) | (.621) | (.685) | (.769) | (.833) | (.892) |
| $\rho_{23}$ | $\rho_{12}$ | $\rho_{13}$ | | | | | | | | |
| .1 | .4 | .4 | .943 | .912 | .886 | .861 | .841 | .813 | .768 | .763 |
| .1 | .5 | .4 | .962 | .935 | .908 | .882 | .860 | .829 | .801 | .772 |
| .1 | .5 | .5 | .978 | .954 | .929 | .902 | .880 | .846 | .815 | .782 |
| | | | β(.277) | (.412) | (.518) | (.601) | (.689) | (.761) | (.828) | (.889) |
| .2 | .4 | .4 | .937 | .907 | .881 | .859 | .833 | .810 | .787 | .762 |
| .2 | .5 | .4 | .957 | .929 | .902 | .879 | .852 | .826 | .799 | .771 |
| .2 | .5 | .5 | .973 | .948 | .923 | .899 | .870 | .842 | .812 | .780 |
| | | | β(.258) | (.393) | (.502) | (.595) | (.678) | (.754) | (.823) | (.886) |
| .3 | .4 | .4 | .932 | .902 | .876 | .853 | .830 | .800 | .785 | .761 |
| .3 | .5 | .4 | .952 | .923 | .897 | .873 | .848 | .823 | .797 | .770 |
| .3 | .5 | .5 | .968 | .942 | .917 | .892 | .865 | .838 | .810 | .779 |

However, there are situations where a combination is quite reasonable. For example, if our concern was only with two structural components on a missile we form a new variable

$$U = \frac{A}{\sigma_{x_2}}(x_2 - \mu_{x_2}) + \frac{B}{\sigma_{x_3}}(x_3 - \mu_{x_3})$$

where $(x_2, x_3)$ are measurements on the two structural components with means $(\mu_{x_2}, \mu_{x_3})$ and standard deviations $(\sigma_{x_2}, \sigma_{x_3})$ and where

$$A = \frac{\rho_{12} - \rho_{13}\,\rho_{23}}{\sqrt{1-\rho_{23}^2}\sqrt{1-\rho_{23}^2-\Delta}}, B = \frac{\rho_{13} - \rho_{12}\,\rho_{23}}{\sqrt{1-\rho_{23}^2}\sqrt{1-\rho_{23}^2-\Delta}}$$

and

$$\Delta = 1 - \rho_{12}^2 - \rho_{13}^2 - \rho_{23}^2 + 2\rho_{12}\rho_{13}\rho_{23}.$$

Then the correlation between the performance variable (say, $Y$) and $U$ is given by

$$\rho_{YU} = \frac{\sqrt{1-\rho_{23}^2-\Delta}}{\sqrt{1-\rho_{23}^2}}$$

In other words, it is possible to add measurements of several variables as indicated above and reduce everything to a single screening variables and a single performance variable. This is a viable approach as long as there are not separate minimum requirements for the screening variables. For more information on this see Thomas, Owen and Gunst (1977).

We now will confine the rest of our discussion to a single screening variable ($X$) and a single performance variable ($Y$).

## 2 CASE WHERE ALL PARAMETERS ARE KNOWN

We assume that we have a bivariate normal distribution with a performance variable $Y$ (say, lifetime) and a lower specification limit on $Y$, which we will designate $L$. The proportion of the total population of lifetimes ($Y$) greater than $L$ is designated $\gamma$. We propose to screen on the correlated variable $X$ (say, voltage) so that we raise the proportion of $Y$'s greater than $L$ to $\delta$, i.e., in mathematical terms.

$$P\{Y \geqslant L\} = \gamma, \text{ and}$$

$$P\{Y \geqslant L \mid X \geqslant \mu_x - K_\beta \sigma_x\} = \delta,$$

where $X$ and $Y$ have a joint bivariate normal distribution with positive correlation, $\rho$.

The mean and standard deviation of $X$ are $\mu_x$ and $\sigma_x$, respectively, and $K_\beta$ is a standardized normal deviate corresponding to $100\beta\%$ of a standardized normal distribution in the lower tail of the normal distribution.

Table 3 gives what amounts to the inverse of Table 1. That is, here we set a goal that the proportion $\delta$ must meet the requirements on the performance variable and we tabulate the proportion $\beta$ which must be selected using the screening variable to reach our goal. Table 3 is representative of much more extensive tables which are available in Odeh and Owen (1980).

For example, if we wanted to raise the proportion of acceptable items from 0.75 to 0.95 and the correlation $\rho$ is 0.90 then we would select the upper 68.82% of the $X$ measurements, i.e., select all $X \geqslant \mu_x - 0.4989\sigma_x$.

In our example the original population can be divided into 4 parts:

(1) Those which are accepted by Screening and meet Specifications, i.e., $P\{Y \geqslant L \text{ and } X \geqslant \mu_x - K_\beta\sigma_x\} = \delta\beta = 0.654$.

(2) Those which are rejected by Screening but meet Specifications, i.e., $P\{Y \geqslant L \text{ and } X > \mu_x - K_\beta\sigma_x\} = \gamma - \delta\beta = 0.096$. Note that these two add to $\gamma = P\{Y \geqslant L\}$.

(3) Those which are accepted by Screening but fail Specifications, i.e., $P\{Y \geqslant L \text{ and } X \geqslant \mu_x - K_\beta\sigma_x\} = \beta - \delta\beta = 0.034$.

(4) Those which are rejected by Screening and fail Specifications, i.e., $P\{Y < L \text{ and } X < \mu_x - K_\beta\sigma_x\} = 1 - \gamma - \beta + \delta\beta = 0.216$. Note that these four proportions add to one.

**Table 3 Table of values of $\beta$ where proportion acceptable after selection = $\delta$ = .950**

| Proportion Acceptable in Non-Selected Population $=\gamma$ | Correlation = $\rho$ | | | | | | |
|---|---|---|---|---|---|---|---|
| | .600 | .700 | .750 | .800 | .900 | .950 | 1.000 |
| .750 | .2812 | .4282 | .4989 | .5661 | .6882 | .7432 | .7895 |
| .760 | .3035 | .4509 | .5206 | .5863 | .7043 | .7567 | .8000 |
| .770 | .3273 | .4745 | .5430 | .6070 | .7205 | .7703 | .8105 |
| .780 | .3527 | .4991 | .5661 | .6281 | .7368 | .7839 | .8211 |
| .790 | .3798 | .5246 | .5898 | .6496 | .7532 | .7975 | .8316 |
| .800 | .4086 | .5511 | .6142 | .6715 | .7696 | .8110 | .8421 |
| .810 | .4392 | .5785 | .6392 | .6938 | .7862 | .8246 | .8526 |
| .820 | .4716 | .6069 | .6648 | .7165 | .8028 | .8381 | .8632 |
| .830 | .5060 | .6362 | .6910 | .7395 | .8194 | .8516 | .8737 |
| .840 | .5423 | .6664 | .7178 | .7628 | .8360 | .8651 | .8842 |
| .850 | .5806 | .6975 | .7450 | .7863 | .8526 | .8784 | .8947 |
| .860 | .6209 | .7293 | .7727 | .8100 | .8691 | .8917 | .9053 |
| .870 | .6629 | .7618 | .8007 | .8338 | .8855 | .9048 | .9158 |
| .880 | .7067 | .7947 | .8288 | .8575 | .9016 | .9177 | .9263 |
| .890 | .7519 | .8280 | .8569 | .8811 | .9176 | .9305 | .9368 |
| .900 | .7981 | .8612 | .8848 | .9043 | .9331 | .9430 | .9474 |
| .910 | .8446 | .8940 | .9121 | .9268 | .9482 | .9552 | .9579 |
| .920 | .8905 | .9256 | .9382 | .9484 | .9627 | .9671 | .9684 |
| .930 | .9341 | .9552 | .9626 | .9684 | .9763 | .9785 | .9789 |
| .940 | .9727 | .9811 | .9840 | .9861 | .9889 | .9894 | .9895 |

This population is then divided into two populations, one of which is accepted by screening:

Those which meet Specifications are

$$P\{Y \geqslant L \text{ given } X \geqslant \mu_x - K_\beta \sigma_x\} = \delta = 0.95.$$

Those which fail Specifications are

$$P\{Y < L \text{ given } X \geqslant \mu_x - K_\beta \sigma_x\} = 1 - \delta = 0.05.$$

And in the population which is rejected by Screening;
Those which meet Specifications are

$$P\{Y \geqslant L \text{ given } X < \mu_x - K_\beta \sigma_x\} = \frac{\gamma - \delta\beta}{1 - \beta} = 0.309.$$

Those which fail Specifications are

$$P\{Y < L \text{ given } X < \mu_x - K_\beta \sigma_x\} = \frac{1 - \gamma - \beta + \delta\beta}{1 - \beta} = 0.691$$

In the original population 25% fail to meet specifications while in the population selected by screening only 5% fail to meet specifications. On the other hand in the rejected group 30.9% do meet specifications.

Now it might be well to digress to remark that we have assumed that we had a lower specification limit and a positive correlation.

The procedure is also applicable for the situations where $X$ and $Y$ are negatively correlated or if there is an upper specification limit, $U$, on $Y$. In particular these situations would be handled as follows:

1. Negative correlation and upper specification limit $U$ on $Y$:

   a. enter the table with the absolute value of correlation,
   b. accept all units whose value $X$ exceeds $\mu - K_\beta \sigma$, and
   c. reject all other submitted units.

2. Negative correlation and lower specification limit $L$ on $Y$:

   a. enter Table 3 with the absolute value of the correlation,
   b. accept all units whose value $X$ is less than $\mu + K_\beta \sigma$, and
   c. reject all other submitted units.

3. Positive correlation and upper specification limit $U$ on $Y$:

   a. accept all units whose value $X$ is less than $\mu + K_\beta \sigma$, and
   b. reject all other submitted units.

## 3. CASE WHERE ALL PARAMETERS ARE UNKNOWN

In most applications the parameters of the distributions will be unknown and they will have to be replaced by estimates obtained from a preliminary sample of size $n$ (a training set). The steps in the screening procedure will now be given. These steps require several tables to be entered to obtain the mathematical quantities required for the procedure. Odeh and Owen (1980) give tables for each of these quantities.

(1) A preliminary sample of size $n$ is obtained of paired values $(x_1, y_1) \ldots (x_n, y_n)$ and the usual estimators of the parameters are computed.

(2) A lower $100\eta\%$ confidence limit on $\rho$ is computed and called $\rho^*$. If this is positive, we proceed to step (3). If it is negative an upper $100\eta\%$ limit is also computed. If this is positive, the process stops since the two variables $X$ & $Y$ could then be independent. If the $100\eta\%$ upper confidence limit on $\rho$ is also negative, then we proceed, making modifications indicated above in Section 2 for a negative correlation.

(3) A $100\eta\%$ lower confidence limit on $\gamma = P\{Y \geqslant L\}$ is computed and labeled $\gamma^*$.

(4) Enter a table of the normal conditioned on $T$ distribution with parameters (and estimates) and degrees of freedom $= n - 1$,

$$\gamma^*,$$

$$\rho^* \sqrt{\frac{n}{n+1}},$$

$$\delta.$$

This table is like Table 3 except that it also varies with degrees of freedom.

(5) All product items are accepted if

$$X \geqslant \bar{x} - t_\beta \sqrt{\frac{n+1}{n}} s_x$$

(6) We can then be at least $100(2\eta - 1\%$ sure that at least $100\delta\%$ of the $Y$'s are above $L$ in the selected population.

For example, if a preliminary sample of size 17 is taken and $r = 0.94$ then choosing $\eta = .95$ we obtain a 95% lower confidence limit on $\rho$ to be $\rho^* = 0.8558$.

If $k = (\bar{y} - L)/s_y = 2.0$ then a 95% lower confidence limit on $\gamma$ is $\gamma^* = 0.90$.

We enter the normal conditioned on $t$-table with (16,0.90, 0.8317,0.95) for $(f, \gamma, \rho, \delta)$ and obtain $t_\beta = 1.384$. Our criterion is to select all items for which $X \geqslant \bar{x} - 1.424 s_x$. Then in the selected group we can be at least 90% sure that at least 95% of the performance variable, $Y$, is greater than the lower specification limit $L$.

If this screening is performed on a finite group of say $M$, items then the items in that group follow a binomial distribution with parameters $M$ and $\delta$. The situation is very similar to what is called prediction intervals in the literature, except that we say we are at least 100 ($2\eta$ − 1%) sure that the probability of $z$ or less defectives is *at least* that given by the binomial distribution. Hence, if $M = 10$ for the example above with an $\eta = .95$, then we are at least 90% sure that the probability of zero defectives in this group is 0.5999. See the paper by Owen, Li and Chou (1981) for more details on this.

## 4. TWO-SIDED SPECIFICATION LIMITS, KNOWN PARAMETERS

Now let us consider extensions of these procedures to two-sided limits, i.e., where we are interested in controlling the $P\{L \leqslant Y \leqslant U\}$. Here things become much more complicated and it is convenient to define a $\gamma_1$ and $\gamma_2$ which are the $\gamma$'s of the one-sided procedure, i.e., let

$$P\{Y \geqslant L\} = \gamma_1 \qquad L = \mu_\gamma - K_{\gamma_1}\sigma_\gamma$$
$$P\{Y \leqslant U\} = \gamma_2 \qquad U = \mu_\gamma + K_{\gamma_2}\sigma_y.$$

Before screening, the proportion of acceptable product is $\gamma_1 + \gamma_2 - 1$. After screening, $\gamma_1 + \gamma_2 - 1$ is raised to $\delta$.

Because of the many combinations of $\gamma_1$ and $\gamma_2$ possible we sought a solution first when

$$\gamma_1 = \gamma_2 = \gamma.$$

In this case we accept all items for which

$$\mu_x - K_\beta \sigma_x \leqslant X \leqslant \mu_x + K_\beta \sigma_x$$

where $K_\beta$ is read from tables obtained from solving

$$BVN(K_\gamma, K_\beta) - BVN(K_\gamma, -K_\beta) - \frac{\delta + 1}{2}(2\beta - 1) = 0,$$

where BVN is the bivariate normal cumulative with the variates standardized.

Table 4 gives values of $K_\beta$ that solves this equation when $\delta = 0.90$. More extensive tables may be found in Li and Owen (1979).

Note that in the one-sided case we can select so few product that we can raise $\delta$ to any preassigned value. That is, there is no mathematical limit on $\delta$; although there are clearly practical ones depending upon how much good pro-

**Table 4 Table of $K_\beta$ for two-sided specifications with $\gamma_1 = \gamma_2 = \gamma$ and proportion acceptable after selection of $\delta = 0.90$**

| Gamma | Correlation | | |
|---|---|---|---|
| | .90 | .95 | 1.00 |
| .78 | .3373 | .6657 | .8820 |
| .79 | .4368 | .7231 | .9239 |
| .80 | .5252 | .7814 | .9674 |
| .81 | .6085 | .8411 | 1.0129 |
| .82 | .6894 | .9026 | 1.0606 |
| .83 | .7699 | .9664 | 1.1108 |
| .84 | .8517 | 1.0329 | 1.1639 |
| .85 | .9357 | 1.1029 | 1.2206 |
| .86 | 1.0234 | 1.1771 | 1.2816 |
| .87 | 1.1159 | 1.2565 | 1.3476 |
| .88 | 1.2151 | 1.3424 | 1.4202 |
| .89 | 1.3231 | 1.4370 | 1.5011 |
| .90 | 1.4432 | 1.5428 | 1.5932 |
| .91 | 1.5804 | 1.6644 | 1.7013 |
| .92 | 1.7438 | 1.8102 | 1.8339 |
| .93 | 1.9522 | 1.9983 | 2.0099 |
| .94 | 2.2622 | 2.2842 | 2.2865 |

duct you are willing to screen out. For the two-sided case the largest value that $\delta$ can attain is

$$2G\left(\frac{K_\gamma}{\sqrt{1-\rho^2}}\right) - 1.$$

When $\gamma_1 \neq \gamma_2$ we accept all items for which

$$\mu_x - K_{\beta_1}\sigma_x \leqslant X \leqslant \mu_x + K_{\beta_2}\sigma_x$$

where $K_{\beta_1}$, $K_{\beta_2}$ are obtained from tables in equal-tailed cases. The proportion of $Y$'s meeting the specification is

$$\Delta = P\{-K_{\gamma_1} \leqslant Z_1 \leqslant K_{\gamma_2} | -K_{\beta_1} \leqslant Z_2 \leqslant K_{\beta_2}\}$$

$$= \delta - \frac{1}{\beta_1 + \beta_2 - 1} \int_{K_{\gamma_1}}^{K_{\gamma_2}} \int_{-K_{\beta_2}}^{-K_{\beta_1}} g(u, v)\, du\, dv$$

$$= \delta - \text{adj. } \delta,$$

where $g(u, v)$ is the density function for a standardized bivariate normal distribution.

We then computed values of this adjustment to $\delta$ and obtained the results given in Table 5.

**Table 5 Values of maximum adjustments to $\delta$ when using equal tail specifications in unequal tail cases**

| $\delta = 0.90$ | adj. $\delta$ is largest when $\rho = 0.70$ | | |
|---|---|---|---|
| | $\gamma_1$ | $\gamma_2$ | adj. $\delta$ |
| | .88 | .89 | .004309 |
| | .88 | .90 | .006953 |
| | .88 | .91 | .008708 |
| | .88 | .92 | .009912 |
| | .88 | .93 | .010772 |
| | .88 | .94 | .011446 |
| | all other combinations of $\rho$ and $\gamma_1$, $\gamma_2$ | | |
| | max. adj. $\delta = 0.006$ | | |
| $\delta = 0.95$ | max. adj. $\delta = 0.006$ | | |
| $\delta = 0.99$ | max. adj. $\delta = 0.001$ | | |

Now let us illustrate the procedure by an example:

$Y$ = voltage at an internal point of a device

$X$ = voltage at an external point of a device

$L = 12$ volts $\quad U = 16$ volts

$\mu_y = 13.8$ volts $\quad \sigma_y = 2.13$ volts $\quad \rho = 0.90$

Then,

$$K_{\gamma_1} = 0.845 \qquad K_{\gamma_2} = 1.033$$

or

$$P\{Y \geqslant L\} = \gamma_1 = 0.80, \quad P\{Y \leqslant U\} = \gamma_2 = 0.85$$

For $\delta = 0.90$, we get $K_{\beta_1} = 0.5252$, $K_{\beta_2} = 0.9357$ and we will accept all items for which

$$\mu_x - 0.5252\sigma_x \leqslant X \leqslant \mu_x + 0.9357\sigma_x$$

In the selected group, at least 89.4% of the $Y$ values will be between 12 volts and 16 volts.

## 5. TWO SIDED SPECIFICATION LIMITS AND UNKNOWN PARAMETERS

When parameters are unknown we start with a training set of size $n$ and proceed through the following four steps:

1. Find $\gamma_1^*$, $\gamma_2^*$ such that

$$P\{\gamma_1 \geqslant \gamma_1^*, \gamma_2 \geqslant \gamma_2^*\} = \eta$$

where $\gamma_1^*$, $\gamma_2^*$ can be obtained from a table of a bivariate noncentral $t$-distribution. See Owen (1965) for a short table of this and see Owen and Frawley (1971) for a larger table.

2. Find $\rho^*$ such that

$$P\{\rho \geqslant \rho^*\} = \eta,$$

Odeh and Owen (1980) give a table for obtaining this easily.

3. If $\sigma_x$ is known, enter table with $\left(\rho^* \sqrt{\frac{n}{n+1}}, \delta, \gamma_1^*\right)$ to obtain $K_{1_{\beta_1^*}}$ and similarly, to obtain $K_{2_{\beta_2^*}}$. Compute $K_{\beta_1^*} = \sqrt{\frac{n+1}{n}} K_{1_{\beta_1^*}}$, $K_{\beta_2^*} = \sqrt{\frac{n+1}{n}} K_{2_{\beta_2^*}}$. If $\sigma_x$ is unknown, tables have not yet been prepared.

4. Select all $X$ for which

$$\bar{x} - K_{\beta_1^*}\sigma_x \leqslant X \leqslant \bar{x} + K_{\beta_2^*}\sigma_x.$$

Then, in the selected population we will be at least $100(2\eta - 1)\%$ confident that approximately $100\delta\%$ of the $Y$'s are between $L$ and $U$.

## 6. CONCLUSION

The screening procedures given here are all based on a bivariate normal model. If such a model does not obtain, the solution seems to lie in the direction of transformations to bivariate normality. A comprehensive approach to such transformations is needed.

A great deal of work will probably be required for each application in devising the variable or variables to be used as screening variables. This also is a topic for further work.

## REFERENCES

Li, L. and D.B. Owen (1978). Two-sided Screening Procedures in the Bivariate Case. *Technometrics*, Vol. 20, 79-85.

Odeh, Robert E. and D.B. Owen (1980). *Tables for Normal Tolerance Limits, Sampling Plans and Screening.* New York: Marcel Dekker, Inc.

Owen, D.B. (1965). A special case of a bivariate noncentral $t$-distribution. *Biometrika*, Vol. 52, 437-446.

Owen, D.B. and J.W. Boddie (1976). A screening method for increasing acceptable product with some parameters unknown. *Technometrics*, Vol. 18, 195-199.

Owen, D.B. and W.H. Frawley (1971). Factors for tolerance limits which control both tails of the normal distribution. *Journal of Quality Technology*. Vol. 3, 69-79.

Owen, D.B. and R.W. Haas (1978). Tables of the Normal Conditioned on *t*-distribution. *Contributions to Survey Sampling and Applied Statistics*, edited by H. A. David. Academic Press, New York.

Owen, D.B. and T.S. Hua (1977). Tables of Confidence Limits on the Tail Area of the Normal Distribution. *Communications In Statistics*, Vol. B6, 285-311.

Owen, D.B. and F. Ju (1977). The Normal Conditioned on *t*-distribution when the Correlation is one. *Communications In Statistics*, Vol. B6, 167-179.

Owen, D.B. and L. Li (1980). The Use of Cutting Scores in Selection Procedures. *Journal of Educational Statistics*, Vol. 5, 157-168.

Owen, D.B., L. Li and Y.-M. Chou (1981). Prediction Intervals for Screening Using a Measured Correlated Variate. *Technometrics*, Vol. 23, 165-170.

Owen, D.B., D. McIntire, and E. Seymour (1975). Tables Using One or Two Screening Variables to Increase Acceptable Product Under One-Sided Specifications. *Journal of Quality Technology*, Vol. 7, 127-138.

Owen, D.B. and Y.H. Su (1977). Screening Based on Normal Variables. *Technometrics*, Vol. 19, 65-68.

Packman, R.F., S.J. Klima, R.L. Davies, J. Malpani, J. Mayzis, W. Walker, B.G.W. Yee, and D.P. Johnson (1976). Reliability of flow detection by nondestructive testing, *Metals Handbook*, Vol. 11, 414-424.

Thomas, J.G., D.B. Owen, and R.F. Gunst (1977). Improving the Use of Educational Tests as evaluation Tools. *Journal of Educational Statistics*, Vol. 2, 55-77.

# Faulty Inspection Distributions—Some Generalizations

Norman L. Johnson

*University of North Carolina at Chapel Hill*

Samuel Kotz

*University of Maryland, College Park*

**Abstract**

In Johnson, Kotz and Sorkin (1980), the authors derived the distribution of the number of items *observed* to be defective in samples from a finite population, when detection is *erroneous* with a nonzero probability.

We extend here the above results by taking into account incorrect identification of nondefectives as well as defectives. Corresponding waiting time distributions are also derived. Furthermore, the case of a stratified finite population corresponding, for example, to defective features of differing severity is considered. Numerical values illustrating the dependence of the corresponding probabilities on the two "misidentification" parameters are presented.

## 1 INTRODUCTION

Johnson *et al.* (1980) have discussed some problems arising when attributes inspection is "less than perfect." They considered, in effect, sampling without replacement from a lot of size $N$ containing $X$ defective (or nonconforming) items, when inspection detects such items with probability $p$. Here this is extended (i) by allowing for a probability, $p'$, of erroneously deciding that an item is defective when really it is not and (ii) by stratifying the population so that inspection error probabilities vary from stratum to stratum.

## 2. TWO KINDS OF INSPECTION ERROR

The number of defective items, $Y$, in a random sample (without replacement) of size $n$ has a hypergeometric distribution with parameters $n$, $X$, $N$. Conditionally on $Y$, the number of items *correctly* called "defective" is distributed as binomial $(Y, p)$ and the number *incorrectly* called "defective" is distributed as binomial $(n - Y, p')$. Thus, the overall distribution of the total number of items called "defective," $Z$ say, is

$$\text{Bin}\,(Y, p) + \text{Bin}\,(n - Y, p') \underset{Y}{\wedge} \text{Hypg}\,(n, X, N)$$

(the two binomial variables being mutually independent), where $\wedge$ denotes the compounding operator (Johnson and Kotz (1969, p. 184)).

Conditional on $Y$, the $r^{\text{th}}$ factorial moment of $Z$ is

$$\mu_{(r)}(Z|Y) = E[Z^{(r)}|Y] = \sum_{j=0}^{r} \binom{r}{j} Y^{(j)} p^j (n - Y)^{(r-j)} p'^{r-j}.$$

The unconditional $r^{\text{th}}$ factorial moment of $Z$ is

$$\mu_{(r)}(Z) = \sum_{j=0}^{r} \binom{r}{j} p^j p'^{r-j} E[Y^{(j)}(n - Y)^{(r-j)}]$$

$$= \frac{n^{(r)}}{N^{(r)}} \sum_{j=0}^{r} \binom{r}{j} p^j p'^{r-j} X^{(j)} (N - X)^{(r-j)}. \tag{1}$$

In particular,

$$E[Z] = \{Xp + (N - X)p'\}n/N = n\bar{p} \tag{2.1}$$

$$\text{Var}\,(Z) = n\bar{p}(1 - \bar{p}) - \frac{n(n-1)}{(N-1)} \frac{X}{N} \left(1 - \frac{X}{N}\right) (p - p')^2, \tag{2.2}$$

where $\bar{p} = N^{-1}\{Xp + (N - X)p'\}$. The conditional probability mass function (pmf) of $Z$ given $Y$ is

$$Pr[Z = z|Y] = \sum_{j=0}^{z} \binom{Y}{j} p^j (1 - p)^{Y-j} \binom{n - Y}{z - j} p'^{z-j} (1 - p')^{n-Y-z+j} \tag{3}$$

$$(z = 0, 1, \ldots, Y),$$

where $\binom{a}{b} = 0$ if $a < b$.

Hence the unconditional pmf of $Z$ is

$$Pr[Z = z] =$$

$$\binom{N}{n}^{-1} \sum_{y} \binom{X}{y} \binom{N - X}{n - y} \sum_{j=0}^{z} \binom{y}{j} \binom{n - y}{z - j} p^j (1 - p)^{y-j} p'^{z-j} (1 - p')^{n-y-z+j}, \tag{4}$$

when the first $\Sigma$ is over $\max\,(0, n - N + X) \leqslant y \leqslant \min\,(n, X)$.

**Table 1**

$p = 0.75$

| $z/p'$ | N = 100; X = 5 | | | | | N = 200; X = 10 | | | | |
|---|---|---|---|---|---|---|---|---|---|---|
| | 0 | 0.025 | 0.05 | 0.075 | 0.1 | 0 | 0.025 | 0.05 | 0.075 | 0.1 |
| 0 | .6731 | .5244 | .4060 | .3122 | .2384 | .6778 | .5280 | .4087 | .3142 | .2399 |
| 1 | .2818 | .3575 | .3891 | .3901 | .3711 | .2736 | .3520 | .3855 | .3879 | .3698 |
| 2 | .0422 | .1010 | .1608 | .2139 | .2557 | .0446 | .1015 | .1603 | .2129 | .2545 |
| 3 | .0028 | .0156 | .0379 | .0679 | .1028 | .0038 | .0167 | .0388 | .0684 | .1030 |
| 4 | .0001 | .0015 | .0056 | .0138 | .0267 | .0002 | .0017 | .0060 | .0143 | .0272 |
| 5 | — | .0001 | .0006 | .0019 | .0047 | — | .0001 | .0006 | .0020 | .0049 |
| 6 | — | — | — | .0002 | .0006 | — | — | — | .0002 | .0006 |
| 7 | — | — | — | — | — | — | — | — | — | .0001 |
| 8 | — | — | — | — | — | — | — | — | — | — |
| 9 | — | — | — | — | — | — | — | — | — | — |
| 10 | — | — | — | — | — | — | — | — | — | — |
| | N = 100; X = 10 | | | | | N = 200; X = 20 | | | | |
| 0 | .4459 | .3488 | .2711 | .2094 | .1606 | .4524 | .3537 | .2749 | .2122 | .1627 |
| 1 | .3892 | .3985 | .3867 | .3611 | .3274 | .3803 | .3928 | .3831 | .3591 | .3265 |
| 2 | .1369 | .1920 | .2378 | .2720 | .2940 | .1363 | .1902 | .2354 | .2695 | .2917 |
| 3 | .0252 | .0513 | .0831 | .1180 | .1532 | .0274 | .0529 | .0840 | .1182 | .1529 |
| 4 | .0027 | .0084 | .0183 | .0327 | .0513 | .0034 | .0093 | .0193 | .0336 | .0521 |
| 5 | .0002 | .0009 | .0027 | .0060 | .0116 | .0003 | .0011 | .0030 | .0064 | .0120 |
| 6 | — | .0001 | .0003 | .0008 | .0018 | — | .0001 | .0003 | .0008 | .0019 |
| 7 | — | — | — | .0001 | .0002 | — | — | — | .0001 | .0002 |
| 8 | — | — | — | — | — | — | — | — | — | — |
| 9 | — | — | — | — | — | — | — | — | — | — |
| 10 | — | — | — | — | — | — | — | — | — | — |
| | N = 100; X = 20 | | | | | N = 200; X = 40 | | | | |
| 0 | .1856 | .1465 | .1149 | .0896 | .0694 | .1913 | .1509 | .1183 | .0922 | .0714 |
| 1 | .3540 | .3209 | .2860 | .2510 | .2173 | .3507 | .3193 | .2856 | .2514 | .2181 |
| 2 | .2870 | .3028 | .3093 | .3077 | .2992 | .2813 | .2977 | .3051 | .3043 | .2966 |
| 3 | .1299 | .1619 | .1915 | .2174 | .2386 | .1299 | .1609 | .1899 | .2154 | .2364 |
| 4 | .0362 | .0542 | .0751 | .0980 | .1221 | .0382 | .0559 | .0762 | .0987 | .1223 |
| 5 | .0065 | .0119 | .0195 | .0295 | .0419 | .0075 | .0130 | .0206 | .0306 | .0429 |
| 6 | .0008 | .0017 | .0034 | .0060 | .0098 | .0010 | .0021 | .0038 | .0065 | .0103 |
| 7 | .0001 | .0002 | .0004 | .0008 | .0015 | .0001 | .0002 | .0005 | .0009 | .0017 |
| 8 | — | — | — | .0001 | .0002 | — | — | — | .0001 | .0002 |
| 9 | — | — | — | — | — | — | — | — | — | — |
| 10 | — | — | — | — | — | — | — | — | — | — |

**Table 1** *(continued)*

$p = 0.8$

| $z/p'$ | 0 | 0.025 | 0.05 | 0.075 | 0.1 | 0 | 0.025 | 0.05 | 0.075 | 0.1 |
|---|---|---|---|---|---|---|---|---|---|---|
| | $N = 100;\ X = 5$ | | | | | $N = 200;\ X = 10$ | | | | |
| 0 | .6545 | .5096 | .3942 | .3029 | .2311 | .6598 | .5136 | .3973 | .3052 | .2328 |
| 1 | .2946 | .3642 | .3915 | .3896 | .3687 | .2855 | .3581 | .3875 | .3871 | .3673 |
| 2 | .0474 | .1073 | .1670 | .2192 | .2597 | .0499 | .1077 | .1664 | .2180 | .2584 |
| 3 | .0034 | .0172 | .0405 | .0712 | .1065 | .0046 | .0184 | .0414 | .0718 | .1068 |
| 4 | .0001 | .0017 | .0062 | .0148 | .0282 | .0002 | .0020 | .0066 | .0153 | .0287 |
| 5 | — | .0001 | .0006 | .0021 | .0050 | — | .0001 | .0007 | .0022 | .0053 |
| 6 | — | — | — | .0002 | .0006 | — | — | .0001 | .0002 | .0007 |
| 7 | — | — | — | — | .0001 | — | — | — | — | .0001 |
| 8 | — | — | — | — | — | — | — | — | — | — |
| 9 | — | — | — | — | — | — | — | — | — | — |
| 10 | — | — | — | — | — | — | — | — | — | — |
| | $N = 100;\ X = 10$ | | | | | $N = 200;\ X = 20$ | | | | |
| 0 | .4206 | .3285 | .2549 | .1966 | .1505 | .4276 | .3339 | .2591 | .1997 | .1529 |
| 1 | .3960 | .3991 | .3830 | .3547 | .3196 | .3866 | .3930 | .3793 | .3527 | .3186 |
| 2 | .1500 | .2038 | .2473 | .2788 | .2980 | .1490 | .2014 | .2444 | .2759 | .2954 |
| 3 | .0297 | .0575 | .0905 | .1258 | .1609 | .0321 | .0592 | .0913 | .1260 | .1605 |
| 4 | .0034 | .0099 | .0208 | .0362 | .0557 | .0043 | .0110 | .0219 | .0372 | .0566 |
| 5 | .0002 | .0011 | .0031 | .0069 | .0130 | .0004 | .0014 | .0035 | .0074 | .0135 |
| 6 | — | .0001 | .0003 | .0009 | .0020 | — | .0001 | .0004 | .0010 | .0022 |
| 7 | — | — | — | .0001 | .0002 | — | — | — | .0001 | .0002 |
| 8 | — | — | — | — | — | — | — | — | — | — |
| 9 | — | — | — | — | — | — | — | — | — | — |
| 10 | — | — | — | — | — | — | — | — | — | — |
| | $N = 100;\ X = 20$ | | | | | $N = 200;\ X = 40$ | | | | |
| 0 | .1632 | .1284 | .1004 | .0781 | .0603 | .1691 | .1330 | .1039 | .0807 | .0623 |
| 1 | .3385 | .3038 | .2683 | .2337 | .2009 | .3359 | .3028 | .2684 | .2344 | .2020 |
| 2 | .2980 | .3089 | .3109 | .3054 | .2937 | .2916 | .3033 | .3063 | .3018 | .2910 |
| 3 | .1463 | .1775 | .2056 | .2295 | .2482 | .1457 | .1760 | .2035 | .2269 | .2455 |
| 4 | .0442 | .0638 | .0860 | .1098 | .1343 | .0464 | .0655 | .0871 | .1103 | .1343 |
| 5 | .0086 | .0150 | .0237 | .0349 | .0486 | .0098 | .0163 | .0251 | .0362 | .0498 |
| 6 | .0011 | .0023 | .0044 | .0075 | .0119 | .0014 | .0028 | .0049 | .0082 | .0127 |
| 7 | .0001 | .0002 | .0005 | .0011 | .0020 | .0001 | .0003 | .0007 | .0012 | .0022 |
| 8 | — | — | — | .0001 | .0002 | — | — | .0001 | .0001 | .0002 |
| 9 | — | — | — | — | — | — | — | — | — | — |
| 10 | — | — | — | — | — | — | — | — | — | — |

**Table 1** *(continued)*

$p = 0.85$

| | $N = 100;\ X = 5$ | | | | | $N = 200;\ X = 10$ | | | | |
|---|---|---|---|---|---|---|---|---|---|---|
| $z/p'$ | 0 | 0.025 | 0.05 | 0.075 | 0.1 | 0 | 0.025 | 0.05 | 0.075 | 0.1 |
| 0 | .6363 | .4950 | .3827 | .2938 | .2240 | .6421 | .4995 | .3861 | .2964 | .2259 |
| 1 | .3068 | .3705 | .3936 | .3889 | .3661 | .2968 | .3637 | .3892 | .3862 | .3646 |
| 2 | .0528 | .1136 | .1732 | .2244 | .2637 | .0553 | .1140 | .1724 | .2230 | .2621 |
| 3 | .0040 | .0189 | .0431 | .0746 | .1103 | .0054 | .0203 | .0442 | .0753 | .1106 |
| 4 | .0001 | .0019 | .0067 | .0158 | .0298 | .0003 | .0023 | .0073 | .0164 | .0303 |
| 5 | — | .0001 | .0007 | .0023 | .0054 | — | .0002 | .0008 | .0024 | .0057 |
| 6 | — | — | — | .0002 | .0007 | — | — | .0001 | .0002 | .0007 |
| 7 | — | — | — | — | .0001 | — | — | — | — | .0001 |
| 8 | — | — | — | — | — | — | — | — | — | — |
| 9 | — | — | — | — | — | — | — | — | — | — |
| 10 | — | — | — | — | — | — | — | — | — | — |
| | $N = 100;\ X = 10$ | | | | | $N = 200;\ X = 20$ | | | | |
| 0 | .3964 | .3091 | .2396 | .1844 | .1410 | .4040 | .3150 | .2441 | .1879 | .1436 |
| 1 | .4012 | .3985 | .3785 | .3479 | .3115 | .3914 | .3922 | .3747 | .3459 | .3106 |
| 2 | .1632 | .2153 | .2563 | .2851 | .3015 | .1615 | .2123 | .2529 | .2817 | .2986 |
| 3 | .0347 | .0641 | .0980 | .1337 | .1686 | .0373 | .0658 | .0988 | .1338 | .1680 |
| 4 | .0042 | .0116 | .0235 | .0398 | .0603 | .0053 | .0129 | .0248 | .0410 | .0613 |
| 5 | .0003 | .0013 | .0037 | .0079 | .0145 | .0005 | .0017 | .0042 | .0085 | .0151 |
| 6 | — | .0001 | .0004 | .0010 | .0024 | — | .0001 | .0005 | .0012 | .0026 |
| 7 | — | — | — | .0001 | .0003 | — | — | — | .0001 | .0003 |
| 8 | — | — | — | — | — | — | — | — | — | — |
| 9 | — | — | — | — | — | — | — | — | — | — |
| 10 | — | — | — | — | — | — | — | — | — | — |
| | $N = 100;\ X = 20$ | | | | | $N = 200;\ X = 40$ | | | | |
| 0 | .1432 | .1123 | .0876 | .0678 | .0522 | .1492 | .1170 | .0912 | .0706 | .0543 |
| 1 | .3219 | .2861 | .2507 | .2166 | .1849 | .3199 | .2857 | .2512 | .2178 | .1863 |
| 2 | .3066 | .3127 | .3105 | .3014 | .2868 | .2996 | .3067 | .3056 | .2977 | .2841 |
| 3 | .1626 | .1927 | .2190 | .2405 | .2565 | .1613 | .1905 | .2162 | .2373 | .2533 |
| 4 | .0530 | .0741 | .0974 | .1219 | .1466 | .0553 | .0758 | .0984 | .1222 | .1463 |
| 5 | .0110 | .0186 | .0285 | .0410 | .0559 | .0126 | .0202 | .0301 | .0425 | .0572 |
| 6 | .0015 | .0031 | .0056 | .0093 | .0144 | .0019 | .0036 | .0063 | .0101 | .0153 |
| 7 | .0001 | .0003 | .0007 | .0014 | .0025 | .0002 | .0004 | .0009 | .0016 | .0028 |
| 8 | — | — | .0001 | .0001 | .0003 | — | — | .0001 | .0002 | .0003 |
| 9 | — | — | — | — | — | — | — | — | — | — |
| 10 | — | — | — | — | — | — | — | — | — | — |

**Table 1** *(continued)*

$p = 0.9$

| $z/p'$ | N = 100; X = 5: 0 | 0.025 | 0.05 | 0.075 | 0.1 | N = 200; X = 10: 0 | 0.025 | 0.05 | 0.075 | 0.1 |
|---|---|---|---|---|---|---|---|---|---|---|
| 0 | .6184 | .4808 | .3714 | .2849 | .2170 | .6249 | .4858 | .3752 | .2878 | .2192 |
| 1 | .3183 | .3762 | .3953 | .3879 | .3634 | .3074 | .3689 | .3905 | .3849 | .3618 |
| 2 | .0584 | .1201 | .1793 | .2296 | .2675 | .0610 | .1203 | .1783 | .2279 | .2657 |
| 3 | .0047 | .0206 | .0458 | .0780 | .1141 | .0063 | .0222 | .0470 | .0787 | .1144 |
| 4 | .0002 | .0021 | .0073 | .0169 | .0313 | .0004 | .0026 | .0080 | .0176 | .0320 |
| 5 | — | .0001 | .0008 | .0024 | .0058 | — | .0002 | .0009 | .0027 | .0061 |
| 6 | — | — | .0001 | .0002 | .0007 | — | — | .0001 | .0003 | .0008 |
| 7 | — | — | — | — | .0001 | — | — | — | — | .0001 |
| 8 | — | — | — | — | — | — | — | — | — | — |
| 9 | — | — | — | — | — | — | — | — | — | — |
| 10 | — | — | — | — | — | — | — | — | — | — |
| | N = 100; X = 10 | | | | | N = 200; X = 20 | | | | |
| 0 | .3733 | .2907 | .2250 | .1729 | .1319 | .3815 | .2971 | .2298 | .1766 | .1347 |
| 1 | .4049 | .3968 | .3734 | .3407 | .3032 | .3947 | .3904 | .3696 | .3287 | .3024 |
| 2 | .1762 | .2263 | .2648 | .2908 | .3045 | .1738 | .2227 | .2609 | .2870 | .3012 |
| 3 | .0400 | .0709 | .1057 | .1417 | .1762 | .0428 | .0726 | .1064 | .1415 | .1754 |
| 4 | .0052 | .0135 | .0263 | .0437 | .0651 | .0065 | .0150 | .0278 | .0450 | .0661 |
| 5 | .0004 | .0016 | .0043 | .0089 | .0161 | .0006 | .0020 | .0049 | .0096 | .0169 |
| 6 | — | .0001 | .0005 | .0012 | .0027 | — | .0002 | .0006 | .0014 | .0030 |
| 7 | — | — | — | .0001 | .0003 | — | — | — | .0001 | .0003 |
| 8 | — | — | — | — | — | — | — | — | — | — |
| 9 | — | — | — | — | — | — | — | — | — | — |
| 10 | — | — | — | — | — | — | — | — | — | — |
| | N = 100; X = 20 | | | | | N = 200; X = 40 | | | | |
| 0 | .1253 | .0980 | .0761 | .0587 | .0450 | .1314 | .1027 | .0798 | .0615 | .0471 |
| 1 | .3044 | .2682 | .2331 | .1999 | .1695 | .3031 | .2684 | .2341 | .2015 | .1713 |
| 2 | .3128 | .3142 | .3081 | .2957 | .2786 | .3052 | .3079 | .3031 | .2920 | .2760 |
| 3 | .1787 | .2073 | .2314 | .2503 | .2635 | .1766 | .2043 | .2279 | .2465 | .2598 |
| 4 | .0626 | .0851 | .1093 | .1342 | .1588 | .0649 | .0867 | .1101 | .1343 | .1582 |
| 5 | .0140 | .0227 | .0339 | .0477 | .0638 | .0159 | .0246 | .0358 | .0493 | .0652 |
| 6 | .0020 | .0040 | .0070 | .0114 | .0173 | .0026 | .0047 | .0079 | .0124 | .0184 |
| 7 | .0002 | .0005 | .0010 | .0018 | .0031 | .0003 | .0006 | .0012 | .0021 | .0035 |
| 8 | — | — | .0001 | .0002 | .0004 | — | — | .0001 | .0002 | .0004 |
| 9 | — | — | — | — | — | — | — | — | — | — |
| 10 | — | — | — | — | — | — | — | — | — | — |

**Table 1** *(continued)*

| | $p = 0.95$ | | | | | | | | | |
|---|---|---|---|---|---|---|---|---|---|---|
| | $N = 100;\ X = 5$ | | | | | $N = 200;\ X = 10$ | | | | |
| $z/p'$ | 0 | 0.025 | 0.05 | 0.075 | 0.1 | 0 | 0.025 | 0.05 | 0.075 | 0.1 |
| 0 | .6184 | .4808 | .3714 | .2849 | .2170 | .6249 | .4858 | .3752 | .2878 | .2192 |
| 1 | .3183 | .3762 | .3953 | .3879 | .3634 | .3074 | .3689 | .3905 | .3849 | .3618 |
| 2 | .0584 | .1201 | .1793 | .2296 | .2675 | .0610 | .1203 | .1783 | .2279 | .2657 |
| 3 | .0047 | .0206 | .0458 | .0780 | .1141 | .0063 | .0222 | .0470 | .0787 | .1144 |
| 4 | .0002 | .0021 | .0073 | .0169 | .0313 | .0004 | .0026 | .0080 | .0176 | .0320 |
| 5 | — | .0001 | .0008 | .0024 | .0058 | — | .0002 | .0009 | .0027 | .0061 |
| 6 | — | — | .0001 | .0002 | .0007 | — | — | .0001 | .0003 | .0008 |
| 7 | — | — | — | — | .0001 | — | — | — | — | .0001 |
| 8 | — | — | — | — | — | — | — | — | — | — |
| 9 | — | — | — | — | — | — | — | — | — | — |
| 10 | — | — | — | — | — | — | — | — | — | — |
| | $N = 100;\ X = 10$ | | | | | $N = 200;\ X = 20$ | | | | |
| 0 | .3733 | .2907 | .2250 | .1729 | .1319 | .3815 | .2971 | .2298 | .1766 | .1347 |
| 1 | .4049 | .3968 | .3734 | .3407 | .3032 | .3947 | .3904 | .3696 | .3287 | .3024 |
| 2 | .1762 | .2263 | .2648 | .2908 | .3045 | .1738 | .2227 | .2609 | .2870 | .3012 |
| 3 | .0400 | .0709 | .1057 | .1417 | .1762 | .0428 | .0726 | .1064 | .1415 | .1754 |
| 4 | .0052 | .0135 | .0263 | .0437 | .0651 | .0065 | .0150 | .0278 | .0450 | .0661 |
| 5 | .0004 | .0016 | .0043 | .0089 | .0161 | .0006 | .0020 | .0049 | .0096 | .0169 |
| 6 | — | .0001 | .0005 | .0012 | .0027 | — | .0002 | .0006 | .0014 | .0030 |
| 7 | — | — | — | .0001 | .0003 | — | — | — | .0001 | .0003 |
| 8 | — | — | — | — | — | — | — | — | — | — |
| 9 | — | — | — | — | — | — | — | — | — | — |
| 10 | — | — | — | — | — | — | — | — | — | — |
| | $N = 100;\ X = 20$ | | | | | $N = 200;\ X = 40$ | | | | |
| 0 | .1253 | .0980 | .0761 | .0587 | .0450 | .1314 | .1027 | .0798 | .0615 | .0471 |
| 1 | .3044 | .2682 | .2331 | .1999 | .1695 | .3031 | .2684 | .2341 | .2015 | .1713 |
| 2 | .3128 | .3142 | .3081 | .2957 | .2786 | .3052 | .3079 | .3031 | .2920 | .2760 |
| 3 | .1787 | .2073 | .2314 | .2503 | .2635 | .1766 | .2043 | .2279 | .2465 | .2598 |
| 4 | .0626 | .0851 | .1093 | .1342 | .1588 | .0649 | .0867 | .1101 | .1343 | .1582 |
| 5 | .0140 | .0227 | .0339 | .0477 | .0638 | .0159 | .0246 | .0358 | .0493 | .0652 |
| 6 | .0020 | .0040 | .0070 | .0114 | .0173 | .0026 | .0047 | .0079 | .0124 | .0184 |
| 7 | .0002 | .0005 | .0010 | .0018 | .0031 | .0003 | .0006 | .0012 | .0021 | .0035 |
| 8 | — | — | .0001 | .0002 | .0004 | — | — | .0001 | .0002 | .0004 |
| 9 | — | — | — | — | — | — | — | — | — | — |
| 10 | — | — | — | — | — | — | — | — | — | — |

Numerical values can be obtained expeditiously if adequate tables of hypergeometric probabilities

$$h(y;\ n,\ X,\ N) = \binom{N}{n}^{-1} \binom{X}{y} \binom{N-X}{n-y}$$

and binomial probabilities

$$b(y;\ n,\ p) = \binom{n}{y} p^y (1-p)^{n-y}$$

are available from the formula

$$Pr[Z = z] = \sum_{y} h(y;\ n,\ X,\ N) \sum_{j=0}^{z} b(j;\ y,\ p)\ b\ (z-j;\ n-y,\ p'). \qquad (4)$$

Table 1 gives some examples of distributions, each with sample size $n = 10$. It is limited by space considerations, but fuller tables have been calculated (for other sample sizes as well as other values of $X$ and $N$).

Note that for $p' = 0$, the same distribution is obtained if the values of $n$ and $X$ are interchanged.

As $N$ and $X$ are increased proportionately to each other with $X/N = \omega$, say, the other parameters $(n,\ p,\ p')$ remaining constant, the distribution of $Z$ tends to a binomial with parameters $n,\ XN^{-1}p + (1 - XN^{-1})p'$.

Another simple special case is $p = p'$, leading to a binomial distribution with parameters $n,\ p$ (whatever the values of $X$ and $N$). However, this is a most unlikely situation—it would correspond to completely useless inspection, unable to differentiate between satisfactory and nonsatisfactory items.

The distributions shown are quite sensitive to the value of $p'$ (false condemnation) because the ratio $X/N$ is relatively small. When the proportion of nondefectives $(1 - X/N)$ is lower, $p'$ has less effect.

## 3 STRATIFIED POPULATIONS

More generally, we can suppose the lot divided into $k$ strata $\pi_1, \pi_2, \ldots, \pi_k$ of sizes $N_1, N_2, \ldots, N_k$ $(\sum_{j=1}^{k} N_j = N)$ such that for any chosen individual in $\pi_j$, the probability of "detection as defective" (whether this is really so or not) is $p_j$. The different strata may, for example, correspond to actual defects of differing degrees of visibility. The case considered in Section 2 corresponds to $k = 2$, $p_1 = p,\ p_2 = p',\ N_1 = X,\ N_2 = N - X$.

The number observed as "defective" in a random sample of size $n$ is then distributed as

$$Z \cap \sum_{j=1}^{k} \cap \text{ Bin } (Y_j,\ p_j)_{\underset{\sim}{Y}} \text{ Mult Hypg}_k\ (n;\ N_1,\ \ldots,\ N_k;\ N), \qquad (5)$$

where $\cap$ denotes "distributed as" $\hat{\underset{\sim}{Y}}$. The binomials are mutually independent, conditional on $\underset{\sim}{Y}$; for the multivariate hypergeometric, $Pr[\underset{\sim}{Y} = \underset{\sim}{y}] = \binom{N}{n}^{-1}$

$\prod_{j=1}^{k} \binom{N_j}{y_j}$ $(\Sigma_{j=1}^{k} y_j = n)$. Then

$$E[Z^{(r)} | \underset{\sim}{Y}] = \Sigma'_r \frac{r!}{\prod_{j=1}^{k} r_j!} \prod_{j=1}^{k} (p_j^{r_j} Y_j^{(r_j)}), \tag{6}$$

where $\Sigma'_r$ denotes summation over nonnegative integers $r_1, \ldots, r_k$ such that $\Sigma_{j=1}^{k} r_j = r$. Taking expectations with respect to $\underset{\sim}{Y}$,

$$E[Z^{(r)}] = \Sigma'_r \frac{r!}{\prod_{j=1}^{r} r_j!} \prod_{j=1}^{k} p_j^{r_j} \frac{n^{(r)}}{N^{(r)}} N_j^{(r_j)}$$

$$= \frac{n^{(r)} r!}{N^{(r)}} \Sigma'_r \prod_{j=1}^{k} \left[ \frac{N_j^{(r_j)} p_j^{r_j}}{r_j!} \right]. \tag{7}$$

In particular,

$$E[Z] = \frac{n}{N} \sum_{j=1}^{k} N_j p_j = n\bar{p} \tag{8.1}$$

and

$$E\left[Z(Z-1)\right] = \frac{n(n-1)}{N(N-1)} \left[\sum_{j=1}^{k} N_j(N_j - 1) p_j^2 + 2 \sum\sum_{j<j'}^{k} N_j N_{j'} p_j p_{j'}\right]$$

$$= \frac{n(n-1)}{N(N-1)} \left[\left(\sum_{j=1}^{k} N_j p_j\right)^2 - \sum_{j=1}^{k} N_j p_j^2\right], \tag{8.2}$$

whence

$$\text{Var}(Z) = n\bar{p}(1-\bar{p}) - \frac{n(n-1)}{N(N-1)} \sum_{j=1}^{k} N_j (p_j - \bar{p})^2 \tag{8.3}$$

where $\bar{p} = N^{-1} \Sigma_{j=1}^{k} N_j p_j$.

Alternative and instructive formulae for the variance are

$$\text{Var}(Z) = \frac{n(N-n)}{N-1} \bar{p}(1-\bar{p}) + \frac{n(n-1)}{N(N-1)} \left(\bar{p} - N^{-1} \sum_{j=1}^{k} N_j p_j^2\right) \tag{8.3$'$}$$

$$\text{Var}(Z) = \frac{n(N-n)}{N-1} \bar{p}(1-\bar{p}) + \frac{n(n-1)}{N^2(N-1)} \sum_{j=1}^{k} N_j p_j (1 - p_j). \tag{8.3$''$}$$

The first term in (8.3)′ and (8.3)″ corresponds to the variance of the (actual) number of defectives in a random sample (without replacement) of size $n$ from a population of size $N$ containing $N\bar{p}$ defectives. It follows that the variance of $Z$ is not less than this, while from (8.3) it cannot exceed $n\bar{p}(1-\bar{p})$—the value it would have in sampling with replacement from the same population (when the distribution of $Z$ would be binomial with parameters $n, \bar{p}$). This will, of course, also be good approximation when $N$ is large.

As a limiting case, we might have $N_j = 1$ and $k = N$—that is, each item in the lot would have its own probability $(p_j)$ of being declared defective.

Note that this differs from a model in which there is supposed to be a prior distribution of the probability of being declared defective, and the $p_j$'s are regarded as realized values from this distribution. For this latter model, we reach, in effect, a "with replacement" situation, as distinguished from the "without replacement" model we have considered, with the distribution of $Z$ depending on the specific values of the $p_j$'s.

The "with replacement" model can be regarded as a mixture of "without replacement" models—the latter being conditional on the specific sets of values $p_1, p_2, \ldots, p_N$. The average variance of the number of items declared defective for the "without replacement" will, in general, be smaller than that for the "with replacement"—because the latter is increased by variation among the $p$'s. Formally,

$$\text{Var}(Z) = E_{\underset{\sim}{p}}[\text{Var}(Z|\underset{\sim}{p})] + \text{Var}_{\underset{\sim}{p}}(E[Z|\underset{\sim}{p}]). \tag{9}$$

## 4 A WAITING TIME DISTRIBUTION

Suppose now that, under the conditions of Section 3, items are inspected one at a time (without replacement) until a predetermined number $a$ of items have been assessed as "defective." Denoting by $M$ the number of items needed to attain this goal, we have (cf. Section 4 of Johnson *et al.* (1980))

$$Pr(M > m) = Pr(Z < a) \quad m = 1, 2, \ldots, N-1 \; (a \leqslant N), \tag{10}$$

where $Z$ has the distribution (5) with $n$ replaced by $m$. Note that in this case $\underline{a}$ can exceed the actual number of defectives in the lot because an item can be assessed as defective even though it is not.

The distribution of $M$ is not proper because there is a positive probability that even when all $N$ items in the lot have been inspected, fewer than $\underline{a}$ items will be declared "defective."

In order to derive the distribution of $M$, we first consider the possible sets of decisions for the $N$ items. The distribution of the number $(D)$ of those which will be "defective" is that of the sum of independent binomial variables $B_1, B_2, \ldots, B_k$ with parameters $(N_1, p_1), (N_2, p_2), \ldots, (N_k, p_k)$ respectively. Given $D$, each of the $\binom{N}{D}$ possible orderings of the $D$ "defective" and $(N-D)$ "not defective" decisions is equally likely, and the number of items up to and including the $a^{\text{th}}$ defective $(M)$ has the negative hypergeometric distribution

$$Pr(M=m) = \binom{N}{D}^{-1} \binom{m-1}{a-1} \binom{N-m}{D-a} \quad (m = a, a+1, \ldots, N-D+a), \quad (11)$$

provided $D \geqslant a$.

The conditional expected value of $M$ given $D(\geqslant a)$ is

$$E[M|D] = a(N+1)/(D+1). \quad (12)$$

If we neglect the possibility that $D$ is less than $a$, the overall expected value of $M$ is approximately

$$\begin{aligned} a(N+1)E[(D+1)^{-1}] &\doteqdot [\{E[D]+1\}^{-1} + \{E[D]+1\}^{-3}\,\mathrm{Var}(D)]\,(N+1)a \\ &= a(N+1)\,(\sum_{j=1}^{k} N_j p_j + 1)^{-1} \\ &\quad [1 + (\sum_{j=1}^{k} N_j p_j + 1)^{-2} \sum_{j=1}^{k} N_j p_j (1-p_j)] \\ &\doteqdot a\bar{p}^{-1}[1 + \frac{1}{N}\{1 + \frac{1}{\bar{p}} + \frac{1}{\bar{p}^2} \sum_{j=1}^{k} \omega_j p_j (1-p_j)\}], \end{aligned} \quad (13)$$

where $\omega_j = N_j/N$ (the proportion in the $j^{\text{th}}$ stratum).

We evaluate the overall variance of $M$ as

$$\begin{aligned} \mathrm{Var}\,(M) &= \mathrm{Var}(E[M|D]) + E[\mathrm{Var}(M|D)] \quad (\text{cf. } (9)) \\ &= a^2(N+1)^2\,\mathrm{Var}\,((D+1)^{-2}) + a(N+1)E\left[\frac{(N-D)\,(D+1-a)}{(D+1)^2\,(D+2)}\right]. \end{aligned}$$

After straightforward though quite tedious calculations, we find

$$\begin{aligned} \mathrm{Var}(M) &\doteqdot \frac{a^2}{N\bar{p}^4} \Sigma\omega_j p_j (1-p_j) + \\ &\quad + \frac{a(1-\bar{p})}{\bar{p}^2}\left[1 + \frac{1}{N} + \frac{1}{N\bar{p}}\left\{\frac{3-\bar{p}}{\bar{p}(1-\bar{p})} \sum_{j=1}^{k} \omega_j p_j (1-p_j) - (3+a)\right\}\right] \\ &\doteqdot \frac{a(1-\bar{p})}{\bar{p}^2}\left[1 + \frac{1}{N} - \frac{3+a}{N\bar{p}} + \frac{3+a-(a+1)\bar{p}}{N\bar{p}^2(1-\bar{p})} \sum_{j=1}^{k} \omega_j p_j (1-p_j)\right] \end{aligned} \quad (14)$$

As $N \to \infty$, the expected value of $M$ tends to $a\bar{p}^{-1}$ and the variance tends to $\frac{a}{\bar{p}^2}(1-\bar{p})$. These are the mean and variance, respectively, of the (negative binomial) waiting time distribution for occurrence of $\underline{a}$ "successes" in independent trials with probability of success equal to $\bar{p}$ at each trial.

(14) can be written

Var $(M) \doteq$

$$\frac{a(1-\bar{p})}{\bar{p}^2}\left[1+\frac{1}{N}-\frac{3+a}{N\bar{p}}\left\{1-\frac{\Sigma\omega_j p_j(1-p_j)}{\bar{p}(1-\bar{p})}\right\}-\frac{a+1}{1}\cdot\frac{\Sigma\omega_j p_j(1-p_j)}{\bar{p}(1-\bar{p})}\right].$$

Since $\Sigma_{j=1}^{k}\,\omega_j p_j(1-p_j) = \bar{p} - \Sigma_{j=1}^{k}\,\omega_j p_j^2 \leqslant \bar{p} - \bar{p}^2 = \bar{p}(1-\bar{p})$ and $\dfrac{a+3}{\bar{p}} >$ $a+1$, it follows that

$$\frac{a(1-\bar{p})}{\bar{p}^2}\left[1+\frac{1}{N}-\frac{3+a}{N\bar{p}}\right] \leqslant \operatorname{Var}(M) \leqslant \frac{a(1-\bar{p})}{\bar{p}^2}\left(1+\frac{1}{N}\right).$$

## ACKNOWLEDGEMENT

Samuel Kotz's work was supported by the U.S. Office of Naval Research under Contract N00014-81-K-0301.

## REFERENCES

Johnson, N.L. and Kotz, S. (1969). *Distributions in Statistics—Discrete Distributions.* New York: John Wiley and Sons.

Johnson, N.L. Kotz, S. and Sorkin, H.L. (1980). Faulty inspection distributions. *Comm. Statist. A9,* 917-922.

Sorkin, H.L. (1977). An Empirical Study of Three Confirmation Techniques: Desirability of Expanding the Respondent's Decision Field. Ph.D. Thesis, University of Minnesota.

# Properties of Some Time-Sequential Statistics in Life-Testing†

Joseph C. Gardiner

*Michigan State University*

**Abstract**

For a large class of statistical experiments in connection with longitudinal studies in reliability and life testing the observable variables have a natural time-ordering. We consider a class of time-sequential statistics associated with these ordered variates and derive their distributional properties. In particular we consider the total time under test statistics, progressively censored likelihood ratio statistics and occurrence-exposure rates of multiple decrement theory.

## 1. INTRODUCTION

There is a class of statistical experiments in reliability, life-testing and clinical trials where the observable variables have a natural time-ordering. For example, in a simple life test in which a sample of items are placed simultaneously on test the first observable is the time of failure of the first item to respond followed by the second, and so on, until the largest failure time observed is recorded last. The failure times of the units under investigation are therefore gathered sequentially in increasing order of magnitude. In some situations where failure could be attributed to one of several (but finite number) causes, both the failure times and the decrement type may be recorded. Such experimentation often demands long periods of observation if the complete decrement profiles of the entire sample of specimens under test are to be recorded. A variety of circumstances such as financial and time constraints may preclude the luxury of this prolonged surveillance of experimental units and accordingly

†Research sponsored in part by the Office of Naval Research under ONR Contract N00014-79-C-0522.

in practice various observational plans have been advocated which terminate experimentation before the entire sample has responded. The consideration of such plans is especially cogent in clinical studies involving animal or human subjects. In a toxicological experiment for example, where duration to toxicosis is being measured together with its associated severity, ethical reasons would be compelling enough to force curtailment of investigation when there is strong early evidence of adverse toxicity.

Of the simplest observational designs that have been utilized in practice are Type I censoring (or truncation), where the experiment is monitored for only a preassigned duration, and alternatively, Type II censoring in which experimentation is terminated once a prespecified proportion of units from the sample on test have responded. The truncation time point of Type I censoring and the observed proportion of responses in Type II censoring are prespecified in advance at the onset of experimentation and statistical analysis is performed on the response data collected up to study termination. Evidently both the designs suffer several deficiencies stemming from this ad hoc specification. In the Type I scheme premature truncation typically enhances the risk of erroneous decisions based on sparse information and therefore prolonged experimentation may still be necessary in order to obviate this risk. This must then be assessed against the increased cost and depletion of experimental units. In Type II censoring the time of test termination is random and thus the inherent possibility of protracted on-test time for units may be at variance with other limitations on time and cost.

In recent times more elaborate censoring designs have found favor for use in some longitudinal studies. Some of these combine the more desirable features of the Type I and Type II disciplines. We shall describe here the class of designs called progressively censored schemes (PCS).

The salient characteristic of PCS is that the experiment is monitored from the onset and the information gathered continually revised so that experimentation may be ceased at any stage should it seem warranted by the current accumulated statistical evidence. Therefore in PCS we have the potential of terminating study at the earliest possible stage with the desirable attendant reduction in cost, time and depletion of experimental units. Progressively censored schemes by their constitution are sequential in structure but differ from classical sequential designs in that, in PCS, there is a priori a maximum number of observations that can be made in the experiment. Furthermore, the observable variates in PCS are ordered and therefore nonindependent in general, whereas most classical sequential designs deal with independent variables. This introduces additional complications and subtleties in the analysis of PCS. In common with classical designs however, PCS also contains in its infrastructure considerations of stopping rules, each rule being now indexed by the total number of sample units under consideration.

We now describe the basic features of a multiple decrement model. There are some situations in which 'failure' can be attributed to one of several causes and interest may, for one reason or other, be directed to a particular cause. As an example, consider a complex electronic system whose smooth performance is dependent on the functioning of all of a finite number of components. System failure then may be caused by malfunction of any one of these components and we may be concerned with a particular component whose replacement cost is relatively larger than that of the others. In the case of a laboratory animal experiment concerned with exposure to certain carcinogenic pollutants the terminal response of death or tumorigenesis may be due to one of several organotropic carcinomas (intestine, lung, liver, bladder, etc.).

In this framework we can envisage statistical studies in which a group of experimental units are under observation with both the duration of time to failure and the associated causes are recorded for each unit. However, once again it may become necessary to consider censoring designs which terminate observation before all units have responded. Furthermore, as the occurrence of a failure by one cause necessarily precludes failure by all other risks, investigations focusing on a particular competing risk are complicated by this reduction of the sample units at risk.

## 2. APPLICATIONS

### 2.1 Progressively Censored Likelihood Ratio Statistics

Let us consider a typical life or dosage-response study in which $n \geqslant 1$ individuals are under observation. Denote by $X_1, \ldots, X_n$ the individual response times of the subjects. The observable variables are however the order statistics $X_{n,1}, \ldots, X_{n,n}$ corresponding to $X_1, \ldots, X_n$. If sampling is carried out under the Type I plan where the experiment is terminated after a predetermined duration of time $t \in (0, \infty)$, and if $r^* = r^*(t) \in [0, n]$ responses have been collected in this period, it is reasonable to base statistical investigations on the likelihood corresponding to the observed variates $X_{n,1}, \ldots, X_{n,r^*}$, if $r^* \geqslant 1$; if $r^* = 0$ one uses the likelihood function of none of the $n$ possible responses times being observed prior to time $t$. On the other hand if the experiment is to be monitored up to a prespecified $r$th response (Type II plan) the termination time is $X_{n,r}$ and, as before, inference is based on $X_{n,1}, \ldots, X_{n,r}$. In progressive censoring however, we do not preclude the possibility of test curtailment prior to $X_{n,r}$ through continuous monitoring from the onset. Therefore if for some $k$, $1 \leqslant k \leqslant n$, the statistical information contained in $X_{n,1}, \ldots, X_{n,k}$ warrants a clear decision experimentation is ceased following the $k$th response, that is at time $X_{n,k}$. We note that both $k$, the stopping number and $X_{n,k}$, the stopping time are random variables. Accordingly, we conceive of a family of stopping rules $\{\tau_n: n \geqslant 1\}$ where for each $n \geqslant 1$, $\tau_n \in \{1, \ldots, n\}$ and the event $[\tau_n = k]$ is wholly determined by the observations $X_{n,1}, \ldots, X_{n,k}$. In this situation we wish to describe certain likelihood ratio statistics which have application to asymptotic estimation problems under progressively censoring.

We fix our mathematical framework by specifying a statistical model in which the response time $X$ induces a probability distribution $V_\theta$ on the Borel line $(R, B)$. Here $\theta$ is a parameter lying in some open subset $\Theta$ of $R$. Let $\{X_i;\ i \geqslant 1\}$ be a sequence of independent random variables each $X_i$ have the distribution $V_\theta$ and $P_\theta$ denote the product measure induced by $V_\theta$ on the infinite direct product of copies of $(R, B)$. For each $n \geqslant 1$ and $k$, $1 \leqslant k \leqslant n$, $B_{n,k}$ is the $\sigma$-field generated by $X_{n,1}, \ldots, X_{n,k}$; if $\tau_n$ is a stopping rule adapted to $\{B_{n,k};\ 1 \leqslant k \leqslant n\}$ denote by $P_{\theta,n}$ the projection of $P_\theta$ onto the stopped $\sigma$-

field $B_{n,\tau_n}$. Then if $\{\theta_n\} \in \Theta$ is a sequence of parameter points we define progressively censored likelihood ratio statistics (PCLRS) by $\Lambda(\theta, \theta_n) = dP_{\theta_n,n}/dP_{\theta,n}$ (that is, the Radon-Nikodyn derivative of the absolutely continuous component of $P_{\theta_n,n}$ with respect to $P_{\theta,n}$). Note that when $\tau_n$ is degenerate at $n$, this reduces to the likelihood ratios of the $n$ independent variables $X_1, \ldots, X_n$ induced by distributions $V_{\theta_n}$ and $V_\theta$. Under certain conditions the measures $P_{\theta,n}$ exhibit the property of local asymptotic normality. We shall say the sequence $\{P_{\theta,n}: \theta \in \Theta\}$ is locally asymptotically normal (LAN) at the point $\theta_0 \in \Theta$ if for some real sequence $\{\phi_n: n \geqslant 1\}$ satisfying $0 < \phi_n \uparrow \infty$ as $n \to \infty$, we have a representation

$$\Lambda_n(\theta_0, \theta_0 + \phi_n^{-1}u) = \exp\left\{u\Delta_n - \frac{1}{2}u^2 + \rho_n\right\} \tag{1}$$

where $\{\Delta_n; n \geqslant 1\}$ is asymptotically standard normal under $P_{\theta_0}$ and for each $u \in R$, $\rho_n = \rho_n(u, \theta_0) \to 0$ in $P_{\theta_0}$-probability. There have been in recent times several investigations of the notion of local asymptotic normality of a sequence of probability distributions. The usefulness of this concept in problems of the theory of asymptotic estimation and hypothesis testing has been demonstrated in the works of LeCam and Hajek and in LeCam (1960) a comprehensive examination is made of conditions ensuring the LAN property for different families of distributions. Attention has been originally focused on the distributions induced by a sequence of independent and identically distributed random variables or those connected with a stationary Markov chain. See, for example, Hajek (1972), Roussas (1972) Ibragimov and Khas'minskii (1972), Inagaki and Ogata (1975). In a later paper Ibragimov and Khas'minskii (1975), and Inagaki and Ogata (1977) view the likelihood ratio statistics as a stochastic process in continuous time and then derive some weak convergence results through which the asymptotic properties of maximum likelihood and Bayes estimators are obtained. For the progressively censored framework Sen (1976) makes an initial investigation of the role of statistics $\{\Delta_n: n \geqslant 1\}$ of (1) in hypothesis testing problems followed by derivation of the LAN property and weak convergence results in Gardiner (1981, 1982).

*Example:* Let us suppose that the duration variables $X_i$ have distribution function $F_\theta$ with density $f_\theta$ on the positive part of the real line. Then the joint density of $\underset{\sim}{X}_{n,k} = X_{n,1}, \ldots, X_{n,k}$ is

$$p(\underset{\sim}{x}^{(k)};\theta) = \left(\frac{n!}{(n-k)!}\right)\left(\prod_{i=1}^{k} f_\theta(x_i)\right)(1 - f_\theta(x_k))^{n-k} \tag{2}$$

for $0 < x_1 < \ldots < x_k < \infty$. Fix $\theta_0 \in \Theta$ and consider a sequence $\theta_n = \theta_0 + un^{-1/2} \in \Theta$, $u \in R$. Then with an associated stopping rule $\tau_n$ the PCLRS become

$$\Lambda_n(u) \equiv \Lambda_n(\theta_0, \theta_n) = p(\underline{X}_{n,\tau_n}; \theta_n/p(\underline{X}_{n,\tau_n}; \theta_0), \; n \geqslant 1. \tag{3}$$

To describe the expansion (1) in this case we need more notation. Denote

$$\xi_{n,k} = \left[\frac{\partial}{\partial\theta}(\log p(\underline{X}_{n,k};\ \theta))\right]_{\theta=\theta_0};\ 1 \leqslant k \leqslant n \tag{4}$$

and write

$$\phi_n = \phi_n(\theta_0) = E_{\theta_0}(\xi^2_{n,\tau_n}).$$

Define

$$\Delta_{n,k} = \Delta_{n,k}(\theta_0) = \phi_n^{-1/2}\xi_{n,k};\ 1 \leqslant k \leqslant n. \tag{5}$$

Now set

$$J_\gamma = J_\gamma(\theta_0) = \int_0^{F_{\theta_0}^{-1}(\gamma)} \left[\frac{\partial}{\partial\theta}\left(\log \frac{f_\theta(x)}{1 - F_\theta(x)}\right)\right]^2_{\theta=\theta_0} dF_{\theta_0}(x),\ \gamma \in (0,\ 1]. \tag{6}$$

We need a mild restriction on the growth of $\tau_n$ with $n$. Let $n^{-1}\tau_n \to \gamma \in (0,\ 1]$ in $P_{\theta_0}$-probability. Then under certain regularity conditions we may obtain the expansion

$$\Lambda_n(u) = \exp\left\{uJ_\gamma^{1/2}\ \Delta_{n,\tau_n} - \frac{1}{2}u^2 J_\gamma + \rho_n\right\} \tag{7}$$

where for each $u \in R$, $\rho_n = \rho_n(\theta_0,\ u) \to 0$ in $P_{\theta_0}$-probability and $L(\Delta_{n,\tau_n}|P_{\theta_0}) \to N(0,\ 1)$. The quantity $J_\gamma$ of (6) is related to the Fisher Information Function. Indeed for a unit sample from $F_{\theta_0}$, $J_1$ is this quantity. The sequence $\{\Delta_{n,\tau_n};\ n \geqslant 1\}$ can be employed in the construction of tests in some hypothesis testing problems of reliability. Consider, for example the problem of testing $H_0;\ \theta = \theta_0$ vs $H_1\colon \theta > \theta_0$, $\theta_0$ specified. We have under study $n \geqslant 1$ identical units and desire a procedure of significance level $\alpha \in (0,\ 1)$. One suitable stopping mechanism in this context is connected with the total time on test statistic. We monitor the experiment until the $\tau_n$th failure where

$$\tau_n = \min\{1 \leqslant k \leqslant n\colon \sum_{i=1}^{k} X_{n,i} + (n-k)X_{n,k} > c_n\} \tag{8}$$

and $\{c_n;\ n \geqslant 1\}$ is a sequence of constants to be specified. By selecting $c_n = nc, c = E_{\theta_0}(X_1 \wedge F_{\theta_0}^{-1}(\gamma))$, $\gamma \in (0,]$ it can be shown that $n^{-1}\tau_n \to \gamma$ in $P_{\theta_0}$-probability, (Gardiner (1978)). The test procedure calls for the rejection of $H_0$ if $\xi_{n,\tau_n} > K_n$, where the constants $\{K_n\}$ are obtained, approximately, through (7). Indeed if $z_\alpha$ is the $100(1-\alpha)$-percentage point of the standard normal distribution a simple computation will shown that $K_n$ may be taken as $n^{1/2}J_\gamma^{1/2}z_\alpha$.

Finally for power computations we use the asymptotic distributions of $\Delta_{n,\tau_n}$ under $P_{\theta_n}$. It can be shown that $L(\Delta_{n,\tau_n}|P_{\theta_n}) \to N(u, 1)$.

In the context of our example it is natural to consider estimators of the parameter $\theta$ in terms of the observations $X_{n,1}, \ldots, X_{n,n}$. Let $\Theta_0$ be a compact subset of $\Theta$ and for each $k$, $1 \leqslant k \leqslant n$, we select an estimator $\hat{\theta}_{n,k}$ of $\theta$ by

$$p(\underline{X}_{n,k}; \hat{\theta}_{n,k}) = \sup_{\theta \in \Theta_0} p(\underline{X}_{n,k}; \theta). \tag{9}$$

Then if $\tau_n$ is a stopping time adapted to $\{B_{n,k}: 1 \leqslant k \leqslant n\}$ and for each $\theta \in \Theta_0$, $n^{-1}\tau_n \rightarrow \gamma \in (0,1]$ in $P_\theta$-probability it can be shown that $\{\hat{\theta}_{n,\tau_n}; n \geqslant 1\}$ is a consistent sequence of estimators for $\theta$ and

$$L[n^{1/2}(\hat{\theta}_{n,\tau_n} - \theta) | P_{\theta,n}] \rightarrow N(0, J_\gamma^{-1}(\theta)), \ \theta \in \Theta_0.$$

We finally remark here that the property of local asymptotic normality at $\theta_0$ for the measures $\{P_{\theta,n}: n \geqslant 1\}$ provides the usual asymptotic minimax results for a sequence of estimators $\{T_{n,\tau_n}\}$ of $\theta$. It can be shown that

$$\lim_{\epsilon \rightarrow 0} \underline{\lim}_{n \rightarrow \infty} \sup_{|\theta - \theta_0| < \epsilon} P_{\theta,n}[|n^{1/2}(T_{n,\tau_n} - \theta)| \geqslant \delta] \geqslant 2\Phi(-\delta J_\gamma^{-1/2}(\theta_0)) \tag{10}$$

where $\Phi(\cdot)$ is the standard normal cumulative distribution function.

## 2.1 Occurrence-Exposure Rates in Multiple Risks Theory

We shall consider an experiment in which $n$ identical units are placed on test and observed continuously over the period $(0, t]$. For each specimen of the sample we record either the time of failure and its associated cause or the time $t$ if failure is not observed by time $t$. In order to specify a concrete probabilistic model, let $X$ denote once again the life time of a unit and $J$ label the underlying cause of failure. Therefore $X \wedge t = \min(X, t)$ is the lifetime to which a unit is under risk over the observational period. Suppose $k$ decrement types are operative (numbered 1 through $k$). The subdistribution functions $Q_i$ (crude risks) are defined by

$$Q_i(t) = P[X \leqslant t, J = i], \ t \in [0, \infty), \ 1 \leqslant i \leqslant k \tag{11}$$

through which the survival function $S$ may be expressed, namely $S(t) = P[X > t] = 1 - \sum_{i-1}^{k} Q_i(t)$. We have assumed here the $k$ risks to be mutually exclusive. We may also consider the cause-specific hazard rates $g_i$, $1 \leqslant i \leqslant k$, given by

$$g_i(t) = Q_i'(t)/S(t). \tag{12}$$

Then $g_i(t)\Delta t$ has the interpretation as the probability of failure in $(t, t + \Delta t]$ by cause $i$ given survival from all risks up to time $t$. In recent times considerable attention has been given to estimation of the cumulative hazard $\int_0^t g_i(y)\,dy$ (Aalen (1976, 1978), Fleming (1978)). The reason for this is that this estimation leads to a natural estimator of the survival function $S(t)$. Note that $h(t) = -S'(t)/S(t)$, the overall hazard function, that is $h(t)$ is the instantaneous probability of failure just beyond time $t$ given survival from all risks up to time $t$. Then, by our assumption that the $k$ causes are mutually exclusive and exhaustive, one has

$$h(t) = \sum_{i=1}^{k} g_i(t) \tag{13}$$

and so in conjunction with the relationship between $S$ and the $Q_i$ we have

$$S(t) = \exp\left[-\sum_{i=1}^{k} \int_0^t g_i(y)\,dy\right]. \tag{14}$$

We also have $Q_i(t) = \int_0^t g_i(y)S(y)\,dy$ and thus all entities of interest can be expressed in terms of the hazard rate function $g_i$.

The observables $(X, J)$ are the basic data which suffice for construction of life tables. Several authors (Gail (1975), Elandt-Johnson (1976), Vaeth (1977)) consider hypothetical lifetimes $Z_1, \ldots, Z_k$, one for each cause and then take $X = \min(Z_i;\ 1 \leqslant i \leqslant k)$ and set $J = i$ if and only if $X = Z_i$. For statistical analysis a joint distribution for the $Z_i$ is then assumed. However since $(X, J)$ are the only observables this formulation leads to identifiability problems (Tsiatis (1975)).

Let us consider a concrete estimation problem in our original formulation. We consider the functions $\gamma_i$ defined by

$$\gamma_i(t) = Q_i(t)\Big/\int_0^t S(y)\,dy = \int_0^t g_i(y)S(y)\,dy\Big/\int_0^t S(y)\,dy. \tag{15}$$

Since the expectation of the exposed lifetime ($= X \wedge t$) is $E(X \wedge t) = \int_0^t S(y)\,dy$, we may interpret $\gamma_i(t)$ as the mean (average) decrement risk due to cause $i$ over the duration $(0, t]$. The $\gamma_i$ are weighted averages and appear in diverse disciplines, predominantly in demography where they are termed 'vital rates' (example: disability rates, fertility rates, mortality rates (Hoem (1976))).

In order to estimate the $\gamma_i$ we consider the corresponding occurrence-exposure rates computed from observations made on a sample of $n$ units during $(0, t]$. The occurrence exposure rate (decrement rate, mortality rate) for a risk type is defined as the ratio of the observed frequency of failures among the $n$ units attributed to that risk, to the total lifetime to which all units have been exposed. Thus the $0/e$ rate $\Gamma_{n,i}(t)$ for risk $i$ is

$$\Gamma_{n,i}(t) = \sum_{l=1}^{n} I[X_l \leqslant t,\ J_l = i]\Big/\sum_{l=1}^{n}(X_l \wedge t). \tag{16}$$

Being the ratio of a discrete variate and a continuous variate the distributional properties of the $\Gamma_{n,i}(t)$ are difficult to obtain. However when $n$ grows without bound we have (Gardiner (1981)) interesting results. With $t$ fixed and the $(X_1, J_1), \ldots, (X_n, J_n)$ assumed independent and identically distributed, we have immediately that $\Gamma_{n,i}(t) \to \gamma_i(t)$ a.s. as $n \to \infty$, for each $i$, $1 \leqslant i \leqslant k$. Thus at each $t \in (0, \infty)$, $\Gamma_{n,i}(t)$ estimates $\gamma_i(t)$ consistently. To understand the behavior of the interaction between decrement types we consider the $k$-vector of $0/e$ rates $\underset{\sim}{\Gamma}_n(t) = (\Gamma_{n,1}(t), \ldots, \Gamma_{n,k}(t))$ and the corresponding $\gamma(t) = (\gamma_1(t), \ldots, \gamma_k(t))$. We can show that, under the assumptions of a finite expected lifetime, that is $E(X) < \infty$,

$$||\lambda(\Gamma_n - \gamma|| \equiv \max_{1 \leqslant i \leqslant k} \sup_{(0,\infty)} |\lambda(t)(\Gamma_{n,i}(t) - \gamma_i(t))| \tag{17}$$

converges to zero, a.s.. Here $\lambda(t) = \int_0^t S(y)\,dy$, is introduced since uniform convergence of $\underset{\sim}{\Gamma}_n$ to $\gamma$ cannot be valid in any neighborhood of the origin. (Of course for each $a > 0$ the uniform convergence above is valid, without the factor $\lambda$, on $(a, \infty)$.)

The limiting distribution of $\underset{\sim}{\Gamma}_n(t)$ is $k$-variate normal. It turns out that any two occurrence-exposure rates, say $\Gamma_{n,i}$, $\Gamma_{n,j}$, are in general (asymptotically) correlated. This correlation is positive or negative according as the corresponding cause-specific hazard rate functions $x \to g_i(x)$, $x \to g_i(x)$ are decreasing or increasing in $(0, t]$. When both $g_i$, $g_j$ are constant functions on $(0, t]$, then $\Gamma_{n,i}$, $\Gamma_{n,j}$ are asymptotically independent. The constancy of all hazard rates $g_i$ on $(0, \infty)$ entails a constant overall hazard rate $h$ and so the life time $X$ of the units must be exponential. However it may happen that the $g_i$ are nonconstant functions and yet the $0/e$ rates $\Gamma_{n,i}$ are asymptotically independent.

In some models for survival analysis a proportionality assumption is tenable, that is, there exists constants $c_i > 0$ with $\sum_{i=1}^{k} c_i = 1$ such that $g_i(t) = c_i h(t)$. Then if $X$ has increasing failure rate (IFR) on $(0, t]$ all $g_i$ will be increasing and thus any pair of $0/e$ rates will be negatively correlated. Likewise if $X$ has decreasing failure rate (DFR) all paris of $0/e$ rates are positively correlated. This holds at any time point in $(0, t]$. If $X$ has increasing failure rate average (IFRA) (that is $x \to x^{-1}\int_0^x h$, is increasing on $(0, \infty)$), the $0/e$ rates $\Gamma_{n,i}(\infty)$, $\Gamma_{n,j}(\infty)$ are negatively correlated.

We may also investigate the weak convergence of the process $t \to \underset{\sim}{\Gamma}_n(t)$ (on the function space $D^k[0, \infty)$ or $D^k[0, \infty]$). It can be shown that the scaled process $t \to n^{1/2}\lambda(t)\ (\underset{\sim}{\Gamma}_n(t) - \underset{\sim}{\gamma}(t))$ converges weakly to a Gaussian process whose covariance structure is that a certain (correlated) Brownian Bridges. With this weak convergence established, we may make random time transforms $t \to \tau_n(t)$ and consider $\underset{\sim}{\Gamma}_n(\tau_n(t))$. This will be typically the case when a censoring discipline is operative in our basic observation plan.

Finally a comparison of the $\gamma_i$, say $\gamma_i(s)$, $\gamma_j(t)$, can be undertaken in terms of its empirical measures $\Gamma_{n,i}(s)$, $\Gamma_{n,j}(t)$. Unfortunately the analysis is rather difficult, due to the nature of the limiting distribution of $(\Gamma_{n,i}(s), \Gamma_{n,j}(t))$, unless some simplifications can be made by additional assumptions on the underlying model.

## REFERENCES

Aalen, O. (1976). Nonparametric inference in connection with multiple decrement models. *Scand. J. Statist. 3*, 15-27.

Aalen, O. (1978). Nonparametric estimation of partial transition probabilities in multiple decrement models. *Ann. Statist. 6*, 534-545.

Elandt-Johnson, R. (1976). Conditional failure time distributions under competing risk theory with dependent failure times and proportional hazard rates. *Scand. Act. J.* 37-51.

Fleming, T. (1978). Asymptotic distribution results in competing risks estimation. *Ann. Statist. 6*, 1071-1079.

Gail, M. (1975). A review and critique of some models used in competing risk analysis. Biom. *31*, 209-222.

Gardiner, J.C. and Sen, P.K. (1978). Asymptotic normality of a class of time-sequential statistics and applications. *Comm. Statist. A7*, 4, 373-388.

Gardiner, J.C. (1981). Convergence of progressively censored likelihood ratio processes in lifetesting. *Sankhya.* 43, Series A Pt 1, 37-51.

Gardiner, J.C. (1982). Local asymptotic normality for progressively censored likelihood ratio processes. *J. Multivariate Analy.* In Press. Michigan State University.

Gardiner, J.C. (1982). The asymptotic distribution of mortality rates in competing risks theory. *Scand. J. Statist.* 9, 31-36.

Gardiner, J.C. (1981). Properties of occurrence-exposure rates in multiple decrement theory. RM 405, Dept. of Statist. & Probability, Michigan State University.

Hajek, J. (1972). Local asymptotic minimax and admissibility in estimation. *Proc. 6th Berkeley Sympos. Math. Statist. Probab. 1*, 175-194. Univ. Calif. Press.

Hoem, J.M. (1976). The statistical theory of demographic rates. *Scand. J. Statist. 3*, 169-185.

Ibragimov, I.A. and Khas'minskii, R.Z. (1972). Asymptotic behavior of statistical estimators in the smooth case. I. Study of the likelihood ratio. *Theory Probability Appl. 17*, 445-462.

Ibragimov, I.A. and Khas'minskii, R.Z. (1975). Local asymptotic normality for non-identically distributed observations. *Theory Probability Appl. 20*, 246-260.

Inagaki, N. and Ogata, Y. (1975). The weak convergence of likelihood ratio random fields and its applications. *Ann. Inst. Statist. Math. 27*, 391-419.

Inagaki, N. and Ogata, Y. (1977). The weak convergence of likelihood ratio random fields for Markov observations. *Ann. Inst. Statist. Math. 29*, 165-187.

LeCam, L. (1960). Locally asymptotically normal families of distribution. *Univ. Calif. Publications in Statist. 3*, 37-98.

Roussas, G.G. (1972). Contiguity of Probability Measures. Cambridge: Cambridge Univ. Press.

Sen, P.K. (1976). Weak convergence of progressively censored likelihood ratio statistics and its role in the asymptotic theory of life testing. *Ann. Statist. 4*, 1247-1257.

Tsiatis, A. (1975). A nonidentifiability aspect of the problem of competing risks. *Proc. Nat. Acad. Sci. USA. 72*, 20-22.

Vaeth, M. (1977). A note on the sampling distribution of the maximum likelihood estimators in a competing risks model. *Scand. Actuarial J.* 81-87.

Vaeth, M. (1979). A note on the behavior of occurrence-exposure rates when the survival distribution is not exponential. *Scand. J. Statist. 6*, 77-80.

# Author Index

N



O



P



R



S



T



V



W



Y



Z

# Subject Index